Iterative Splitting Methods for Differential Equations

CHAPMAN & HALL/CRC
Numerical Analysis and Scientific Computing

Aims and scope:
Scientific computing and numerical analysis provide invaluable tools for the sciences and engineering. This series aims to capture new developments and summarize state-of-the-art methods over the whole spectrum of these fields. It will include a broad range of textbooks, monographs, and handbooks. Volumes in theory, including discretisation techniques, numerical algorithms, multiscale techniques, parallel and distributed algorithms, as well as applications of these methods in multi-disciplinary fields, are welcome. The inclusion of concrete real-world examples is highly encouraged. This series is meant to appeal to students and researchers in mathematics, engineering, and computational science.

Proposals for the series should be submitted to one of the series editors above or directly to:
CRC Press, Taylor & Francis Group
4th, Floor, Albert House
1-4 Singer Street
London EC2A 4BQ
UK

Published Titles

Classical and Modern Numerical Analysis: Theory, Methods and Practice
Azmy S. Ackleh, Edward James Allen, Ralph Baker Kearfott,
  and Padmanabhan Seshaiyer

Computational Fluid Dynamics
Frédéric Magoulès

A Concise Introduction to Image Processing using C++
Meiqing Wang and Choi-Hong Lai

Decomposition Methods for Differential Equations:
  Theory and Applications
Juergen Geiser

Discrete Variational Derivative Method: A Structure-Preserving Numerical
  Method for Partial Differential Equations
Daisuke Furihata and Takayasu Matsuo

Grid Resource Management: Toward Virtual and Services Compliant Grid
Computing
Frédéric Magoulès, Thi-Mai-Huong Nguyen, and Lei Yu

Fundamentals of Grid Computing: Theory, Algorithms and Technologies
Frédéric Magoulès

Handbook of Sinc Numerical Methods
Frank Stenger

Introduction to Grid Computing
Frédéric Magoulès, Jie Pan, Kiat-An Tan, and Abhinit Kumar

Iterative Splitting Methods for Differential Equations
Juergen Geiser

Mathematical Objects in C++: Computational Tools in a Unified Object-
Oriented Approach
Yair Shapira

Numerical Linear Approximation in C
Nabih N. Abdelmalek and William A. Malek

Numerical Techniques for Direct and Large-Eddy Simulations
Xi Jiang and Choi-Hong Lai

Parallel Algorithms
Henri Casanova, Arnaud Legrand, and Yves Robert

Parallel Iterative Algorithms: From Sequential to Grid Computing
Jacques M. Bahi, Sylvain Contassot-Vivier, and Raphael Couturier

Iterative Splitting Methods for Differential Equations

Juergen Geiser

CRC Press
Taylor & Francis Group
Boca Raton   London   New York

CRC Press is an imprint of the
Taylor & Francis Group, an **informa** business

A CHAPMAN & HALL BOOK

Chapman & Hall/CRC
Taylor & Francis Group
6000 Broken Sound Parkway NW, Suite 300
Boca Raton, FL 33487-2742

First issued in paperback 2017

© 2011 by Taylor and Francis Group, LLC
Chapman & Hall/CRC is an imprint of Taylor & Francis Group, an Informa business

No claim to original U.S. Government works

ISBN 13: 978-1-138-11190-5 (pbk)
ISBN 13: 978-1-4398-6982-6 (hbk)

Visit the Taylor & Francis Web site at
http://www.taylorandfrancis.com

and the CRC Press Web site at
http://www.crcpress.com

Preface

This book was written at the Department of Mathematics, Humboldt-University of Berlin, Unter den Linden 6, D-10099 Berlin, Germany.

In this monograph, we discuss iterative splitting methods for evolution equations. The evolution equations we consider are systems of parabolic and hyperbolic equations. We focus on convection-diffusion-reaction equations, heat equations, and wave equations, which are used in the modelling of transport-reaction, heat transfer, and elastic wave propagation.

Iterative splitting methods can be classified as decomposition methods in time and space. Here, we propose efficient iterative splitting methods, with respect to the time and space scales of the discretized operators in the equation. The main advantage of our methods is the efficient usage of computational and memory resources because of the simplification that results from decoupling the equations. We also propose higher-order splitting methods, such that the splitting error is negligible compared to the time and space discretization errors. Real-life applications benefit from the iterative splitting methods presented in this book, because the physics is preserved and all operators appear in the decoupled equations, such that each substep presents the solution of the full coupled system.

In the theoretical part of the book (Chapter 2), the main theorems and main results of the stability and consistency analysis are discussed for ordinary differential equations. The extensions (Chapter 3) of the iterative splitting methods to partial differential equations are discussed and enlarged for the theoretical ideas to spatial- and time-dependent differential equations. In the practical part of the book (Chapter 4), the application of the methods to artificial and real-life problems is presented, together with the benefits of equation decomposition. Finally, conclusions and underlying software packages are discussed (Chapter 5, 6).

This work has been accomplished with and supported by many colleagues and co-workers and I would like to thank them all.

First, I am grateful to my colleagues and co-workers at the Humboldt-University of Berlin for their fruitful discussions and ideas. In particular, I wish to thank Prof. C. Carstensen, Prof. T. Friedrich, Prof. A. Griewank and Prof. A. Mielke, PhD student M. Arab and student assistants F. Krien, T. Zacher, and R. Röhle for their support and ideas. I would like to thank L. Noack for programming FIDOS software and J. Gedicke for programming OPERA-Splitt software and their help in the numerical experiments. Further, I would like to thank Prof. G. Tanoglu, Department of Mathematics, IYTE Campus, Urla, Izmir, Turkey and Dr. R. Steijl, School of Engineering, University of Liverpool, UK for their discussions and help in numerical experiments.

Furthermore, I would like to thank the Federal Ministry of Education and Research (BMBF), which support and fund my works.

My special thanks go to my wife Andrea and my daughter Lilli, who have always supported and encouraged me.

Berlin, December 2010   Jürgen Geiser

Contents

List of Figures

List of Tables

Introduction

In this book, we describe the theoretical and practical aspects of iterative splitting methods, used as decomposition methods in time and space for evolution equations. The decomposition methods are discussed with respect to their effectiveness, simplicity, stability, and consistency, and their capability of solving problems in real-life applications.

Theoretical aspects, such as the consistency of the methods in analyzing the local splitting error, are presented as well as techniques used in the underlying proof. The possibility of applying splitting methods to stiff and nonlinear differential equations are discussed as extensions. Often such problems occur in multiphysics or multiscale problems, where different equations or different time scales have to be handled simultaneously.

Some multiphysics problems are presented in their practical aspects. Here different physical behaviors are represented by simpler decoupled models. The models are related to real-life problems and are motivated by original projects studied in the past.

The decomposition is described in relation to the question of how to decouple the problems efficiently without losing their physical correctness. The aim is to solve the simpler parts with respect to their temporal and spatial behaviors, so that the implementation is easier based on the underlying solver methods.

We put forward the following propositions as contributions made in our present work.

- Consistency and stability results for linear operators (given in ordinary differential or spatially discretized partial differential equations),

- Acceleration of solver processes by decoupling of the full equation into simpler equations,

- Effectiveness of decomposed methods with respect to computational time and memory,

- Theory that is based on standard splitting analysis and can be simply extended to more general cases, e.g., time dependent and spatial dependent,

- Embedding of higher-order time discretization methods in decoupled equations,

- Applications in computational sciences, e.g., flow problems, elastic wave propagation, and heat transfer,

- Splitting method can also be used to couple equations, for example partial and ordinary differential equations or continuum and discrete model equations.

The work presented in the next chapters focuses first on the types of multiphysics problems in which our methods are used. We then discuss in detail our decomposition method, followed by an analysis of the stability and consistency of the proposed decomposition methods. Finally, we end with numerical results from our model problems.

Two of our main contributions to decomposition methods are the attainment of higher-order accuracy and the incorporation of coupling techniques in solving real-life problems.

First, we present the modeling of selected multiphysics problems. We describe the underlying characteristics of the various parts of the equations and their spatio-temporal behavior. This knowledge allows us to design special decoupling methods and to respect the underlying conservation of physical properties, see also applications in [124].

In subsequent chapters, we describe the decomposition methods used to decouple equations, which can be analyzed separately. We assume semi-discretization of the evolution equations to give bounded operators. An extension to unbounded operators is also discussed.

As a benefit of analyzing each equation separately, the time and space decomposition methods can be adapted to obtain the best discretization and solver method for each individual decomposed problem. Large problems are reduced to simpler and efficiently solvable partial problems either in space and time.

For the application of decomposition methods, an intensive study of various fields of application is presented. Here, we discuss time and space discretizations as well as the foundation of splitting methods in time, e.g., operator splitting methods, or in space, e.g., Schwarz decomposition methods.

We use the fact that decomposition methods give benefits in decoupling the appropriate operators with respect to their physical and mathematical behavior, see [69], [106]. For the decomposition, we discuss the scales of each operator and collocate the operators with respect to these scales.

By decoupling the equations, the underlying splitting error can be reduced by higher-order splitting methods, see [96]. Furthermore the discretization methods in time and space influence the order of convergence of the underlying schemes and therefore a higher-order method is necessary for decomposition methods and also discretization methods.

Based on this assumption, we propose for time discretizations higher-order Runge-Kutta methods or BDF methods, and for spatial discretizations higher-order finite volume or finite difference methods, as well as mixed discretization methods with embedded local analytical or semi analytical solutions.

To reveal the advantages over classical time-splitting methods, we explain our new proposed iterative operator splitting method, in terms of accuracy and simplicity of implementation in realistic problems. The improvement of such methods as higher-order methods is discussed in the form of the underlying analysis of the iterative operator splitting method.

In the next step, the extension to partial differential equations is discussed as a further application of iterative operator splitting methods. Due to their spatially discretized nature, in looking at ordinary differential equations and assume to obtain boundable operators, we can consider their stability and consistency analysis to be embedded in the theory of the splitting methods of bounded operators.

Therefore, a closed error analysis of iterative operator splitting methods can be presented for the evolution equation with respect to its stiffness, spatial behavior, and nonlinearity. Overall, the new iterative operator splitting methods are excellent decomposition methods for obtaining higher-order methods and can also be extended to nonlinear and spatial splitting methods, see [87], [114] and, [168].

The content of this book is divided into six chapters and covers a wide range of theoretical and practical issues arising from multiphysics and multi-scale problems.

Chapter 1 introduces the basic concepts of this book.

In Chapter 2, we discuss the decomposition and discretization methods. Here we present the main theorems of the analysis to the splitting methods. We start from standard sequential splitting methods to iterative splitting methods and present their underlying consistency and stability results. The derivation to achieve higher-order splitting methods with respect to the underlying splitting steps is presented. We extend the decomposition concepts based on underlying time integration and time discretization methods to apply the iterative schemes to numerical methods. Examples with Runge-Kutta and BDF methods are motivated. As a main goal we embed the iterative splitting method to multistep methods and apply them to systems of ordinary differential equations.

In Chapter 3, we describe the extension of decomposition methods to spatially discretized partial differential equations. The operator splitting methods are discussed with respect to the underlying operators and their restrictions to boundedness. Recent results on the contributions of these methods to stability and consistency analysis are taken into account. Moreover, the theory of iterative operator splitting methods is described and the connection to other splitting methods, such as sequential splitting methods, is presented. Here, we present the flexibility of iterative splitting methods to other classes of problems. The analysis can also be extended to such problems, see for example [92], [93], [105].

In Chapter 4, we present the application of our methods for numerical problems. In the first part of the chapter, we discuss the verification and benchmarks of decomposition methods. Here we discuss all algorithms of the splitting methods presented with their underlying discretization and solver methods. We present step-by-step benchmark problems for ordinary and partial differential equations to see the benefit of the splitting idea. The discussion is held with respect to standard splitting schemes and computational benefits. Extensions to time-dependent and nonlinear problems are given and allow the application of iterative splitting schemes to be widened.

In the second part of the chapter, we present real-life problems in physical and engineering applications. The first real-life problems involve transport-reaction processes in a radioactive waste disposal scenario. The simulations show results on the contamination of the underlying rock. A second problem is given as an elastic wave propagation model, where our schemes are applied to hyperbolic problems. In such a model, we simulate simple earthquakes and obtain results on the strength of the ground forces, see [37] and [38]. In a third application, we deal with a transport and reaction model that simulates the deposition processes in a CVD (chemical vapor deposition) apparatus. Here, we optimize the homogeneous deposition rates on targets, [101]. Our last application deals with coupled Navier-Stokes and Molecular Dynamics. In such applications, we use the iterative splitting scheme as a coupling method. We present the benefits of computations with molecular dynamics extensions of Navier-Stokes equations.

Our results are computed with various software tools including MATLAB$^{\circledR}$, R^{3}T (a software product developed at the University of Heidelberg, Germany), and FIDOS (a software tool based on MATLAB$^{\circledR}$, developed at the Humboldt-University of Berlin, Germany) and a special software code based on coupling MD and Navier-Stokes equations (a software product developed by Dr. Rene Steijl, School of Engineering, University of Liverpool, UK).

In Chapter 5 we summarize the results of the book and propose some future work.

In Chapter 6 we discuss the underlying Software Tools.

The Appendix lists some notation with our abbreviations, a bibliography, and an index.

Outline of the Book

The book is organized in two parts: a theoretical section and an application section. It can be read starting with the theoretical part (Chapter 2) or the application part (Chapter 4). The introductory chapter (Chapter 1) is necessary for an understanding of the basic models and concepts. The extension part (Chapter 3) can be read as a supplement to get an overview of other fields of application. In Figure 1, the outline of the chapters is illustrated graphically.

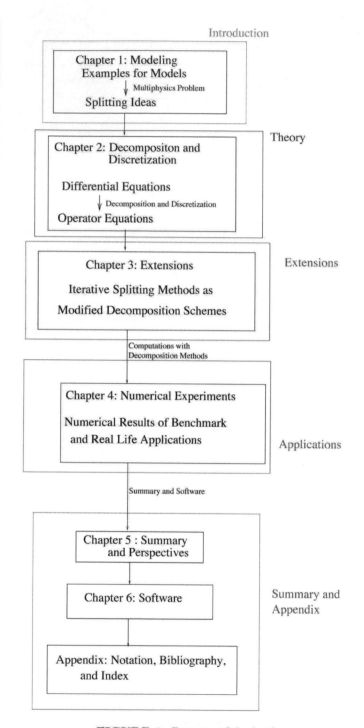

FIGURE 1: Contents of the book.

Chapter 1

Model Problems

In this chapter, we introduce related model problems that can be solved with the splitting methods proposed in our thesis.

In various applications in materials physics, geosciences, and chemical engineering, the simulations of multiphysics problems are very important for predicting quantitative and qualitative behavior and can reduce the number of expensive real-life experiments, see [6] and [162].

For multiphysics problems, we discuss models that are based on different physical processes, e.g., flow, reaction and, retardation processes, which occur over various temporal and spatial scales.

Because of the complicated physical processes and large systems of equations, an enormous increase in the performance of simulation programs is necessary and highly complex algorithms are needed for fast computations.

The mathematical models of multiphysics problems usually consist of coupled differential equations. Furthermore, we deal with ordinary differential equations if we concentrate only on the time variable, e.g., chemical processes, or we extend to partial differential equations if we consider time and spatial variables, e.g., transport processes. At the very least, we describe the underlying physical behavior of the processes with simplified models given in the literature, e.g., [5], [18], and [162]. The most relevant problems cannot be solved analytically and so numerical solutions are necessary.

Based on the physical behavior, the numerical methods can be developed in such a manner that the physical laws are conserved.

We will concentrate on multiphysics problems based on linear and nonlinear systems of coupled parabolic or hyperbolic equations. We assume that if we mix both types (parabolic and hyperbolic) in coupled systems of equations, the systems have characteristics that are at least parabolic.

The questions we address in this thesis about the model problems are:

- How can we decouple the systems of equations into simpler equations while conserving the physical behavior and minimizing the decoupling error ?

- How can we accelerate the solver process for solving the simpler equations ?

- How can we obtain higher accuracy with the splitting methods.

- How can we couple different physical processes while respecting their underlying physical properties, e.g., continuous and discrete problems ?

In the next sections, we discuss some multiphysics problems.

1.1   Related Models for Decomposition

We will concentrate on a family of multiphysics problems related to flow and reaction phenomena. On the one hand, the flow problems can be decoupled into convection, diffusion, and wave-propagation problems, see [17], [18], [27], [41], [46] and [51], while on the other hand, the reaction problems are often ordinary differential equations that can be solved efficiently by explicit or analytical methods, see [122] and [106].

For such relations, we assume at the beginning that we can explicitly consider terms that correspond to flow or reaction processes, see [133] and [149].

Mathematically, we consider a system of equations with operators related to flow or reaction processes. Such a classification helps to produce a decoupled system.

The same idea of finding adequate operators in the model equations is applied when we deal with applications to transport reaction, heat, and flow processes in geophysics, material sciences, and other fields.

The discussion is more abstract if we start with a semi discretized system and end with one operator that contains all the behavior of the equations. Here, we refer to the solver methods, that describe the behavior of eigenvalues, simplified decompositions based on matrices (outer-diagonal and diagonal), see [8], [119], [188], and [146]. Such ideas can help decompose the operators into simpler matrices before applying our proposed splitting methods.

Furthermore, the problem can be discussed for weakly and strongly coupled equations. We assume that strongly coupled equations are systems that can only be solved by increasing the order of the system of equations, otherwise iterative schemes are very slow and decoupling into simpler parts can only be done subject to restrictions, see [24]. Meanwhile weakly coupled equations can be solved and decoupled into simpler equations, see the ideas in [24].

1.2 Examples in Real-Life Applications

The next problems are selected so as to apply our decomposition methods and to illustrate the effect of our equation splitting ideas, see [69], [71], [94], [105], [106], [101], [108], and [109].

In general, the motivation comes from predicting the effect of physical parameters in a model that corresponds to a real-life experiment.

The following points are discussed in the experiments:

- Obtaining parameters of the models used in order to reduce the number of real-life experiments.

- The given model can be solved more simply by using splitting methods.

- Multiphysics problems can be decomposed into simpler physical problems.

The models proposed next are discussed in the literature and numerical results are discussed in Chapter 6.

1.2.1 Waste Disposal

In the first model, we simulate the waste disposal of radioactive waste in a salt dome.

The motivation arose from predicting the possible contamination in a given waste scenario in the surface layer near the disposal area.

Therefore, the following conditions must be fulfilled:

- A model problem must be realistic but solvable as a decomposition problem.

- Results of delicate geometries should be obtained within a week.

- Many simulations with various parameters of diffusion, sorption, etc., should be performed.

Because of real-life experiments and associated laboratory tests being expensive, some geoscientists proposed a simulatable model, which is discussed as follows.

The salt dome is surrounded by a large, heterogeneous overlaying rock layer.

Figure 1.1 presents the physical circumstances of waste disposal as given in the task.

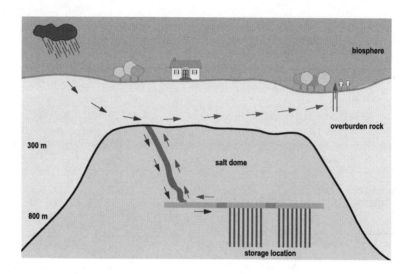

FIGURE 1.1: Schematic overview of waste disposal, cf. [21].

One potential scenario involves pooled groundwater being located in the possible extraction pathway for waste disposal and in contact with the radioactive waste.

The overlaying rock causes high pressure and motion of the salt dome, which in turn expels contaminated water from the salt dome. The contaminated water is then used as a time-dependent source of radionucleides in the groundwater flow.

The radionucleides are transported via groundwater and we neglect the mass of expelled water compared with the water flowing in the groundwater. For reactive processes, we incorporate radioactive decay, which accounts for the transfer of a radionucleide in the corresponding product nucleide, and adsorption, which accounts for exchange in several abidance areas with mobile or immobile phases, see [68].

The modeling for this case is done in [18], [32], [59] and, [150]. Based on the physical model, we derive the mathematical models.

For the sake of simplicity, we describe a model of equilibrium sorption.

This model is studied in various directions and we explain the model in [68]. The phases of the contaminants are presented in the mobile, immobile, sorption, and immobile sorption phases, cf. [68].

The model equations are given by:

$$\phi \, \partial_t \, R_i \, u_i + \nabla \cdot (\mathbf{v} u_i - D \nabla u_i) \quad = \quad -\phi \, R_i \, \lambda_i u_i \tag{1.1}$$

$$+ \sum_{k=k(i)} \phi \, R_k \, \lambda_k u_k + Q_i, \text{ in } \Omega \times [0, T],$$

$$u_{e(i)} \quad = \quad \sum_i u_i, \; R_i = 1 + \frac{(1-\phi)}{\phi} \rho \, K(u_{e(i)}),$$

$$\text{with} \quad i = 1, \ldots, M,$$

where u_i is the i-th concentration, R_i is the i-th retardation factor, λ_i is the i-th decay constant, and $\mathbf{v}$ is the velocity field, which is computed by another program package, e.g., $\mathbf{D^3F}$ program package, see [58], or is given *a priori*. Q_i is the i-th source term, ϕ is the porosity, $c_{e(i)}$ is the sum of all isotope concentrations of the element e, and K is a function of the isotherms, see [68]. M is the number of radioactive contaminants, the domain is given as $\Omega \subset \mathbb{R}^d$, $d = 2, 3$ and $T \in \mathbb{R}^+$ is the end time. Boundary conditions are Neumann and outflow boundary conditions; more details are discussed in Section 6.5 and [68].

Remark 1.1. For this model, we propose the decomposition of flow and reaction processes, which are given on different time scales, and we are free to accelerate the solving process.

1.2.2   Elastic Wave Propagation

In the second model, we simulate the propagation of elastic waves produced by earthquakes.

Here, the motivation arises from predicting possible earthquake formation in a given scenario of earthquakes about the center.

Therefore, the following conditions must be fulfilled:

- A realistic model problem has to consider a singular forcing term.

- Geometries should be in 2D and 3D in order to consider the anisotropic behavior of the earth layers.

- Numerical experiments can be done with physical parameters.

Earthquake simulations motivate the study of elastic wave propagation. Realistic earthquake sources and a complex 3D earth structure are immensely important for understanding earthquake formation. A fundamental model of ground motion in urban sedimentary basins is studied and 3D software packages are developed, see [37] and [38]. Numerical simulations of wave propagation can be performed in two and three dimensions to give models sufficient realism (e.g., three-dimensional geology, propagating sources, and frequencies approaching 1 Hz) to be of engineering interest. Before numerical simulations can be applied in the context of engineering studies or a seismic hazard analysis, the numerical methods and models associated with them must be thoroughly validated.

Here, splitting methods are of enormous interest in order to reduce the computational time and to decouple the problem into simpler and solvable problems.

For such a real-life application, we will consider higher-order splitting methods and perform 3D simulations of simpler earthquake models to test the accuracy of the equation splitting methods as well as the conservation of physical quantities – see the results in Chapter 6.

In the following, we describe the model, which includes the physical parameters of the underlying domains in which the earthquake is formed.

We present the model problem as an elastic wave equation with constant coefficients in the following notation.

$$
\begin{align}
\rho \partial_{tt} U &= \mu \nabla^2 U + (\lambda + \mu) \nabla (\nabla \cdot U) + f, \text{ in } \Omega \times [0, T], \tag{1.2} \\
U(x, 0) &= g_0(x), \text{ on } \Omega, \tag{1.3} \\
\partial_t U(x, 0) &= g_1(x), \text{ on } \Omega, \tag{1.4} \\
U(x, t) &= h(x, t), \text{ on } \partial\Omega \times [0, T], \tag{1.5}
\end{align}
$$

where U is equal to $(u, v)^T$ or $(u, v, w)^T$ in two and three dimensions and f is a forcing function. The initial functions are given as $g_0(x), g_1(x) : \Omega \to \mathbb{R}^d$ and the boundary functions are given as $h(x, t) : \partial\Omega \times [0, T] \to \mathbb{R}^d$. The domain $\Omega \subset \mathbb{R}^d$, with $d = 2, 3$, is sufficiently smooth, see [50]. The material parameters $\lambda, \mu \in \mathbb{R}^+$ reflect the elastic behavior of urban sedimentary basins.

In seismology, it is common practice to use spatially singular forcing terms, which can take the form

$$
f = F\delta(x)g(t), \tag{1.6}
$$

where F is a constant direction vector. A numerical method for solving Equation (1.2) needs to approximate the Dirac function $\delta(x)$ correctly in order to achieve full convergence.

Remark 1.2. Our contributions to this model include the initial process, which is a singular function (Dirac function), and the large time steps in the computation to simulate the necessary time periods. We propose a decomposition with respect to the dimensions (dimensional splitting). Based on the anisotropic parameters of the wave equation, we can accelerate the solving process by using simpler equations and save memory resources for implicit solver methods.

1.2.3 Deposition Models: CVD (Chemical Vapor Deposition) Processes

In this real-life problem, the motivation arose out of developing a new deposition process onto metallic bipolar plates to reduce the process costs and the volume of plates. To realize new deposition processes, Maxphase materials are used, see [12].

The deposition methods are PE-CVD (plasma enhanced chemical vapor deposition) processes, see [162], and [179], for the growth of thin films.

Therefore, the following conditions must be fulfilled:

- Homogenization of deposition layer.

- Deposition onto delicate geometries.

The model is given as follows:

We develop a multiphase model in the following steps, see also [101], [113], [162] and [190]:

- Standard transport model (near field, far field).

- Flow model (flow field of plasma medium).

- Multiphase model with mobile and immobile zones (retardation, adsorption, transport, and reaction processes).

In each part of the model, we can refine the processes of transport of the gaseous deposition or reaction species to reflect the effect of flow field, plasma zones, and precursor gases.

A schematic test geometry of the CVD reactor is given in Figure 1.2.

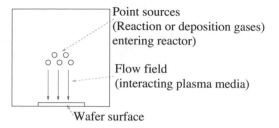

FIGURE 1.2: Vertically impinging CVD reactor.

1.2.3.1   Standard Transport Model

In the following, the models are discussed in terms of far-field and near-field problems, which take into account the different scales of the models.

Two different types of model can be discussed:

1. Convection-diffusion-reaction equations [115] (far-field problem).

2. Boltzmann-Lattice equations [190] (near-field problem).

The modeling is governed by the Knudsen Number, which is a dimensionless number defined as the ratio of molecular mean free path length to representative physical length scale.

$$Kn = \frac{\lambda}{L}, \tag{1.7}$$

where λ is the mean free path and L is the representative physical length scale. This latter length scale could be, for example, the radius of a body in a fluid. Here we deal with small Knudsen Numbers $Kn \approx 0.01 - 1.0$ for a convection-diffusion-reaction equation and a constant velocity field, whereas for large Knudsen Numbers $Kn \geq 1.0$, we deal with a Boltzmann equation [179]. By modeling the gaseous transport of the deposition species, we consider a pure far-field model and assume a continuum flow field, see [113].

Such assumptions lead to transport equations that can be treated by a convection-diffusion-reaction equation with constant velocity field:

$$\partial_t u + \nabla \cdot \mathbf{F} - R_g \;=\; q(x,t), \text{ in } \Omega \times [0,T], \tag{1.8}$$
$$\mathbf{F} \;=\; \mathbf{v}u - D\nabla u,$$
$$u(x,t) \;=\; \tilde{u}_0(x), \text{ in } \Omega, \tag{1.9}$$
$$u(x,t) \;=\; \tilde{u}_1(x,t), \text{ on } \partial\Omega \times [0,T], \tag{1.10}$$

where u is the molar concentration of the reaction gases (called *species*) and

F is the flux of the species. **v** is the flux velocity through the chamber and porous substrate [185]. D is the diffusion matrix and R_g is the reaction term. The initial value is given as $\tilde{u}_0$ and we assume a Dirichlet boundary with a sufficiently smooth function $\tilde{u}_1(x, t)$. $q(x, t)$ is a source function, depending on time and space, and represents the inflow of the species. T is the end time and $\Omega \subset \mathbb{R}^d, d = 2, 3$ is the domain.

The parameters of the equation are derived as follows. Diffusion in the modified CVD process is given by Knudsen diffusion [25]. We consider the overall pressure in the reactor to be 200 Pa and the substrate temperature (or wafer surface temperature) to be about $600 - 900$ $[K]$ (Kelvin). The pore size in the homogeneous substrate is assumed to be 80 $[nm]$ (nanometer). The homogeneous substrate can be either a porous medium, e.g., a ceramic material, see [25], or a dense plasma, assumed to be very dense and stationary, see [162]. For such media we can simulate the diffusion by Knudsen diffusion. The diffusion is described by:

$$D = \frac{2\epsilon\mu_K \mathbf{v} r}{3RT}, \tag{1.11}$$

where ϵ is the porosity, μ_K is the shape factor of Knudsen diffusion, r is the average pore radius, R and T are the gas constant and temperature, respectively, and **v** is the mean molecular speed, given by:

$$\mathbf{v} = \sqrt{\frac{8RT}{\pi W}}, \tag{1.12}$$

where W is the molar mass of the diffusive gas.

For homogeneous reactions, we assume during the CVD process a constant reaction of Si, Ti, and C species given as:

$$3Ti + Si + 2C \rightarrow Ti_3SiC_2, \tag{1.13}$$

where Ti_3SiC_2 is a MAX-phase material, see [12], which is deposited on the wafer surface. For simplicity, we do not consider the intermediate reaction with precursor gases [162] and assume we are dealing with a compound gas $3Ti + Si + 2C$, see [42]. Therefore, we can concentrate on the transport of one species.
The rate of reaction is then given by:

$$\lambda = k_r \frac{[3Si]^M [Ti]^N [2C]^O}{[Si_3TiC_2]^L}, \tag{1.14}$$

where k_r is the apparent reaction constant and L, M, N, O are the reaction orders of the reactants.

A schematic overview of the one-phase model is presented in Figure 1.3. Here the gas chamber of the CVD apparatus is shown, which is modeled by a homogeneous medium.

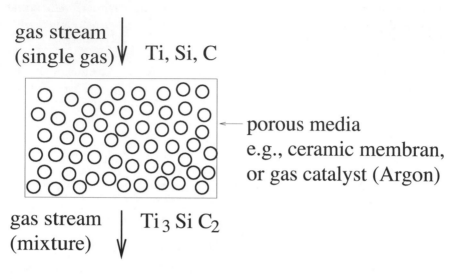

FIGURE 1.3: Gas chamber of CVD apparatus.

1.2.3.2   Flow Field

The flow field for which the velocity is derived is used for the transport of species. The velocity in the homogeneous substrate is modeled by a porous medium [17, 141]. We assume a stationary or low reactivity medium, e.g., non-ionized or low-ionized plasma or less reactive precursor gas. Furthermore, the pressure can be assumed to follow the Maxwell distribution as [162]:

$$p = \rho b T, \tag{1.15}$$

where ρ is the density, b is the Boltzmann constant, and T is the temperature.

The model equations are based on mass and momentum conservation and we assume energy is conserved. Because of the low temperature and low pressure environment, we assume the gaseous flow has a nearly liquid behavior. Therefore, the velocity is given by Darcy's law:

$$\mathbf{v} = -\frac{k}{\mu}(\nabla p - \rho \mathbf{g}), \tag{1.16}$$

where $\mathbf{v}$ is the velocity of the fluid, k is the permeability tensor, μ is the dynamic viscosity, p is the pressure, $\mathbf{g}$ is the vector of gravity, and ρ is the density of the fluid.

We use the continuum equation of particle density and obtain the equation of the system, which is given as a flow equation:

$$\partial_t(\phi\rho) + \nabla \cdot (\rho\mathbf{v}) = Q, \tag{1.17}$$

where ρ is the unknown particle density, ϕ is the effective porosity, and Q is the source term of the fluid.

We assume a stationary fluid and consider only divergence-free velocity fields, i.e.,

$$\nabla \cdot \mathbf{v}(x) = 0, \quad x \in \Omega. \tag{1.18}$$

The boundary conditions for the flow equation are given as:

$$p = p_r(t, \gamma), \quad t \geq 0, \quad \gamma \in \partial\Omega, \tag{1.19}$$

$$\mathbf{n} \cdot \mathbf{v} = m_f(t, \gamma), \quad t \geq 0, \quad \gamma \in \partial\Omega, \tag{1.20}$$

where $\mathbf{n}$ is the normal unit vector with respect to $\partial\Omega$, where we assume that the pressure p_r and flow concentration m_f are prescribed by Dirichlet boundary conditions [141].

From the nearly stationary fluids, we assume that the conservation of momentum for velocity $\mathbf{v}$ is given [114, 141]. Therefore, we can neglect the computation of momentum in the velocity.

Remark 1.3. For flow through a gas chamber, for which we assume a homogeneous medium and non-reactive plasma, we have considered a constant flow [129]. A further simplification is given by the very small porous substrate, for which we assume the underlying velocity in a first approximation to be constant [179].

Remark 1.4. For a non-stationary medium and reactive or ionized plasma, we have to take into account the behavior of the electrons in thermal equilibrium. Such a spatial variation can be considered by modeling electron drift. Such a simulation of the ionized plasma is done using Boltzmanns relation, [162].

1.2.3.3   Multiphase Model: Mobile and Immobile Zones

More complicated processes such as the retardation, adsorption and dissipation processes of the gaseous species are modeled with multiphase equations. We take into account the feature that the concentrations of species can be given in mobile and immobile versions, depending on their different reactive states, see [184]. From these behaviors, we have to model the transport and adsorbed states of species, see also Figure 1.4.

Here, the mobile and immobile phases of the gas concentration are shown on the macroscopic scale of the porous medium.

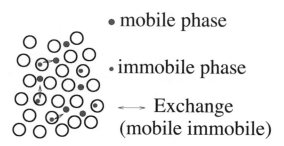

FIGURE 1.4: Mobile and immobile phases.

The model equations are given as combinations of transport and reaction equations (coupled partial and ordinary differential equations) by:

$$\phi\partial_t u_i^L + \nabla \cdot (\mathbf{v}u_i^L - D^{e(i)}\nabla u_i^L) \;=\; g(-u_i^L + u_{i,im}^L) - \lambda_{i,i}\phi u_i^L \qquad (1.21)$$

$$+ \sum_{k=k(i)} \lambda_{i,k}\phi u_k^L + \tilde{Q}_i, \text{ in } \Omega \times [0,T],$$

$$\phi\partial_t u_{i,im}^L \;=\; g(u_i^L - u_{i,im}^L) - \lambda_{i,i}\phi u_{i,im}^L$$

$$+ \sum_{k=k(i)} \lambda_{i,k}\phi u_{k,im}^L$$

$$+\tilde{Q}_{i,im}, \text{ in } \Omega \times [0,T], \qquad (1.22)$$

ϕ :   effective porosity $[-]$,

u_i^L :   concentration of i-th gaseous species in plasma chamber

$u_{i,im}^L$:   concentration of i-th gaseous species in immobile zones

       of plasma chamber

       phase $[mol/mm^3]$,

$\mathbf{v}$:   velocity in plasma chamber $[mm/nsec]$,

$D^{e(i)}$:   element-specific diffusion-dispersion tensor $[mm^2/nsec]$,

$\lambda_{i,i}$:   decay constant of i-th species $[1/nsec]$,

$\tilde{Q}_i$:   source term of i-th species in mobile zones

       of plasma chamber $[mol/(mm^3 nsec)]$,

$\tilde{Q}_{i,im}$:   source term of i-th species in immobile zones

       of plasma chamber $[mol/(mm^3 nsec)]$,

g :   exchange rate between mobile and

       immobile concentrations $[1/nsec]$,

where $i = 1, \ldots, M$ and M denotes the number of components. Further, $1[mm] = 10^{-3}[m]$ means the millimeter scale.

The parameters in Equation (1.21) are further described, see also [68]. The effective porosity is denoted by ϕ and describes the portion of the porosity of the aquifer that is filled with plasma, and we assume a nearly fluid phase. The transport term is indicated by the Darcy velocity $\mathbf{v}$, which presents the flow direction and absolute value of plasma flux. The velocity field is divergence-free. The decay constant of the i-th species is denoted by λ_i. $k(i)$ therefore denotes the indices of the other species. The boundary conditions are given as Neumann and Outflow boundary conditions, see [104] and [108].

Remark 1.5. The concentrations in the mobile zones are modeled by convection-diffusion-reaction equations, see also Section 6.5, whereas the concentrations in the immobile zones are modeled by reaction equations. These two phases represent the mobilities of gaseous species through the homogeneous media, where the concentrations in the immobile zones account at least for the lost amount of depositable gases.

1.2.4   Navier-Stokes Molecular Dynamics: Coupling Continuous and Discrete Problems

In the last model, we widen our focus by applying splitting methods to coupling problems.

In recent years, many applications are based on multiscale problems, when single-scale models, e.g., a continuous model, cannot contain all the information about the problem [6].

Here, we consider multiscale problems in a fluid exhibiting different behavior on macro- and microscales, the model and the underlying computations are discussed in the preprint Geiser/Steijl, see [94].

Such problems arise in micro-electro-mechanical systems (MEMS), which involve flow in micrometer- and nanometer-scale channels [176].

Therefore, the following conditions must be fulfilled:

- The model problem should be computable in the micro model.

- The coupling procedure should be simple to implement.

- Many experiments with various parameters of molecular viscosity should be performed.

In the following, we discuss the solution of time-dependent fluid flows with complex flow phenomena.

In such a computational domain, the Navier-Stokes equations become invalid, e.g., a more detailed computation at the molecular level is needed.

This means that, locally, the flow needs to be modeled in a more detailed physical model.

Here, the Navier-Stokes equations are coupled to a molecular-level simulation based on the methods of Molecular Dynamics.

The Newtonian shear stresses in the Navier-Stokes equations are replaced with the results of Molecular Dynamics simulations.

The Molecular Dynamics method is described in detail in [3], [19], and [197].

For such scale-dependent applications, we apply different spatial meshes for the discretization schemes.

The approach is sketched in Figure 1.5, where the Navier-Stokes equations are discretized on a structured cell-centered finite-volume mesh and the solid blocks in the centers of the cell faces close to the upper and lower boundaries denote microscale problems, which are used to evaluate the shear stresses using Molecular Dynamics.

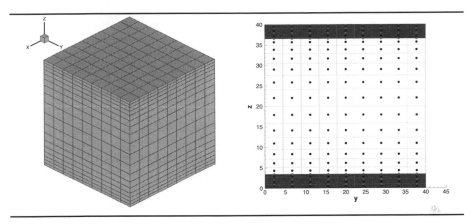

FIGURE 1.5: Cell-centered finite-volume mesh used in channel-flow simulations. A stretched mesh is used in the z-direction (crossflow), while the mesh in the x- and y-directions is uniform. The squares denote the cell faces at which the microscale viscous fluxes are evaluated.

The coupling of the molecular-level simulation to the discretized continuum system is discussed in [193] for steady flows.

The focus of the present work is on time-dependent flows. Hence, the disparity between the underlying timescales in the continuum flow and molecular-level microscale problems forms the main challenge. In the last few years, the solution of multiscale problems has become a major topic. Of particular importance are the solver methods used to couple large scales, see [156] for lattice Boltzmann models.

1.2.4.1 Mathematical Model

Our model equations come from fluid dynamics. The macroscopic equation is given by the Navier-Stokes equation for incompressible continuum flow:

$$\rho \partial_t u + \rho(u \cdot \nabla)u - \mu \Delta u + \nabla p = f, \text{in } \Omega \times (0, T), \qquad (1.23)$$

$$\nabla \cdot u = 0, \text{in } \Omega \times (0, T), \qquad (1.24)$$

$$u(x, 0) = u_0(x), \text{on } \Omega,$$

$$u(x, t) = \mathbf{0}, \text{on } \partial\Omega \times (0, T),$$

The unknown flow vector $u = u(x, t)$ is considered in $\Omega \times (0, T)$. In the above equations, ρ and p represent the fluid density and pressure, respectively. Here,

μ represents the dynamic viscosity of the fluid and is a positive constant. Further $\mathbf{0}$ is the zero vector and $u_0(x)$ is the initial vectorial function. In the momentum equation, Equation (1.23), the term f on the right-hand side represents a volume source term. Equation (1.24) imposes a divergence-free constraint on the velocity field, which is consistent with the assumption of incompressible flow.

The microscopic equation is given by Newton's equation of motion for each individual molecule i in a sample of N molecules,

$$m_i \partial_{tt} x_i = F_i, i = 1, \ldots, N, \tag{1.25}$$

Here, x_i is the position vector of atom i, and the force F_i acting on each molecule is the result of the intermolecular interaction of a molecule i with the neighboring molecules within a finite range of interaction. In the present work, we assume that the interparticle forces are based on the well-known Lennard-Jones interaction potential [159], i.e., we assume that the microscopic flow corresponds to a Lennard-Jones fluid, the details of which are given in a later section.

The coupling between the macroscopic Equation (1.23) and microscopic Equation (6.250) is assumed to take place through the exchange of viscous stresses in the momentum Equation (1.23). The underlying idea is to replace the viscous stresses based on the continuum Newtonian Equation (1.23) with a viscous stress evaluated by Molecular Dynamics simulations of the fluid on the microscale with a velocity gradient on the macroscale imposed on the microscale fluid, via Lees-Edwards boundary conditions [157]. The molecular-level viscosity is evaluated using the Irving-Kirkwood relation [135].

The viscous stress contribution $\mu \Delta u$ in Equation (1.23) can be generalized to the non-Newtonian flow as $\partial \sigma_{ij} / \partial x_j$, using Einstein's summation. In the present work, this non-Newtonian viscous stress contribution is reformulated in the following form:

$$\partial \sigma_{ij} / \partial x_j = \mu_{apparent} \partial^2 u_i / \partial x_j^2 \tag{1.26}$$

where, the "apparent" viscosity can be a general function of the imposed velocity gradients in each spatial direction, i.e., this expression can represent general noncontinuum and non-Newtonian flow conditions. The viscous stresses in Equation (1.23) can now be replaced by molecular-level viscous stresses by introducing a constant approximate viscosity μ_{approx} and by taking into account the deviation of the molecular-level viscous stresses from this approximate viscosity through a volumetric source term.

Thus, finally, we obtain the coupled multiscale equations:

$$\rho \partial_t u + \rho (u \cdot \nabla) u - \mu_{approx} \Delta u + \nabla p = f, \text{ in } \Omega \times (0, T), \tag{1.27}$$
$$f_i = \partial \sigma_{ij} / \partial x_j |_{molecular} - \mu_{approx} \Delta u_i,$$

where Einstein's summation is used for the volumetric source term f, which accounts for the deviation of the viscous stresses evaluated at the molecular level from the approximate Newtonian relation $\mu_{approx}\Delta u$. The molecular-level viscous stresses can be further reformulated using the "apparent" viscosity, as demonstrated in Equation (6.251).

Chapter 2

Iterative Decomposition of Ordinary Differential Equations

In this chapter, we concentrate on the decomposition methods used for ordinary differential equations.

Here, we have the benefit of dealing with bounded operators and can apply the ideas from a large body of literature on general decomposition methods in time and in space, while concentrating on time-decomposition methods. Basic research work was first done on the application to Lie- and Strang splitting methods, see [194] and [201]. Later, various applications to ordinary and also partial differential equations followed, thereby specializing the splitting methods, see [20], [53], [198], etc.

We discuss the ideas of iterative splitting methods. Some background is given on waveform relaxation methods, which have been developed for the iterative solution of differential equations, see [64], [140], and [199]. While the theory goes back to Picard-Lindelöf iterations, see [173], [174], and [175], we give an introduction to iterative splitting methods.

A more applied direction of iterative splitting methods is related to engineering problems. Their origin goes back to a more heuristic concept of the solution of iteratively coupled operators in reducing complicated problems, see the ideas in [86], [147], and [149].

2.1 Historical Overview

The history of iterative time decomposition methods is related to that of iterative solver methods.

We have found two origins of the derivation of splitting schemes:

- Picard-Lindelöf iterations [163, 164], where waveform relaxation methods have their origins [199].

- Iterative split-operator approaches [147], where a two-step method is derived to solve nonlinear reactive transport equations (heuristic approach).

The first origin is more theoretically motivated by solving iterative differential equations [163]. Here the motivation came from applying the algorithms of time and space decomposition methods, see [64], [199].

Here especially waveform relaxation methods are used to solve ordinary and partial differential equations, see [139], [199]. The methods can be accelerated by parallel algorithms and save computational time, e.g., Parareal, see [65].

On the one hand, they help simplify the solver process, while saving time for computing the simple parts of matrices, e.g., diagonal matrices, see [199]. On the other hand, the methods can accelerate the solver process with parallelized algorithms, see [65].

The second origin is motivated more by solving transport reaction problems. Here, the algorithms are focused on solving each part separately to save on computational time, see [147]. Such methods are developed to achieve implementable algorithms and embed standard discretization and solver methods, e.g., concept of sparse matrices, see [87].

From a theoretical point of view, the iterative split-operator algorithm removes the splitting error through iteration and increases the order of convergence with more iteration steps under some assumptions, e.g., bounded operators, see the following sections.

To see the benefit of the second introduction of iterative splitting methods, we concentrate on bounded operators, which are easier to start with than unbounded operators. So we first deal with ordinary differential equations and, in Chapter 5, we extend to partial differential equations, which are semidiscretized. We assume that their underlying spatial discretized operators, which are more or less matrices, can be bound in a Banach space norm, see [86, 107].

From the point of notation, we speak about operator splitting method, if we concentrate on applications to partial differential equations. In a more general case we speak about splitting methods, if we apply to differential equations, means partial differential equations (PDEs), and also ordinary differential equations (ODEs).

In this chapter, we briefly introduce the underlying ideas of the decomposition analysis for bounded operators that are used in ordinary differential equations. Some extensions are done to spatial discretized partial differential equations, which can be studied for our time decomposition methods.

2.2   Decomposition Ideas

In this section, we discuss decomposition ideas used on evolution equations to obtain separate underlying equations.

We assume our discussion is focused on following partial differential equations, with time and space variables. We talk about semidiscretization, when we spatially discretize the differential equation and obtain for simplification ordinary differential equations. In this notation, the operators are matrices, for which we assume a range of bounded operators are derived, see [180].

In our introduction to splitting methods, we concentrate on two ideas for decomposing evolution equations, nevertheless other decompositions can also be found.

The two decomposition ideas are:

- Physical decomposition (related to the underlying processes).

- Mathematical decomposition (related to abstract matrices found in the differential equations after spatial discretization).

Both ideas decouple a given evolution equation into simpler problems with respect to the temporal, spatial, and physical scales of the original equations.

2.2.1 Physical Decomposition

In the physical decomposition, the motivation is to decouple each physical process, which results in a simpler evolution equation.

The more information we have at the beginning of the solver process, the more we can also use the information from the underlying discretization methods. Therefore, we can *a priori* decouple the evolution equations from the underlying methods we used.

Here, the physical decomposition can apply the information on time-scale in different processes. For example, the spatial and time discretizations of convection-diffusion-reaction equations have to be balanced by the Courant-Friedrichs-Levy (CFL) condition, see [71], to achieve accurate results.

In the following, we discuss

- Direct decoupling method;

- Decoupling method based on numerical methods (indirect decoupling);

in our physical decomposition.

2.2.1.1 Direct Decoupling Method

Often, the values of the physical parameters are so obvious, that we can determine the operators of the splitting methods.

The criteria for the direct decoupling method are strong anisotropy in the space dimensions, obvious timescales, e.g., fast reaction process and slow transport process, or different physical scales, e.g., growth and decay processes.

For these obvious problems, we can directly choose the form of the operators, e.g., flow operator and reaction operator.

In an example of the direct decoupling method, we focus on the parabolic differential equation:

$$\frac{\partial u}{\partial t} = \frac{\partial}{\partial x} D_1(x,y) \frac{\partial u}{\partial x}$$

$$+ \frac{\partial}{\partial y} D_2(x,y) \frac{\partial u}{\partial y}, \quad (x,y) \in \Omega, \ t \in [0,T], \tag{2.1}$$

$$u(x,y,0) = u_0(x,y), \quad (x,y) \in \Omega, \tag{2.2}$$

$$u(x,y,t) = g(x,y,t), \quad (x,y) \in \partial\Omega, \ t \in [0,T], \tag{2.3}$$

where the anisotropy of the heat operator is apparent:

$$\max_{(x,y)\in\Omega} D_1(x,y) \ll \min_{(x,y)\in\Omega} D_2(x,y), \tag{2.4}$$

and $D_1, D_2 \in C(\overline{\Omega})$.

Therefore, decoupling in the spatial dimensions is possible, see [126] and [204].

2.2.1.2   Decoupling Method Based on Numerical Methods

The next decoupling possibility we discuss comes after the discretizations in time and space. Based on our discretization methods, stability conditions are necessary to obtain stable solutions, see [182].

Due to this restriction, the influence of physical parameters is important for stability criteria.

One such condition is the so-called *Courant-Friedrichs-Levy (CFL) condition*, see [34], in which no explicit, unconditionally stable, consistent finite difference schemes for hyperbolic initial value problems exist.

So the CFL condition gives the relation between spatial and time discretization with respect to the physical parameters, for which we have unconditionally stable results.

This condition can be used to decouple the equations into different operators, when they present different timescales.

In the example for the decoupling method based on numerical methods, we examine the transport-reaction equation:

$$\frac{\partial u}{\partial t} = v \cdot \nabla u + \lambda u, \quad x \in \Omega, \ t \in [0,T], \tag{2.5}$$

$$u(x,0) = u_0(x), \quad x \in \Omega, \tag{2.6}$$

$$u(x,t) = g(x,t), \quad x \in \partial\Omega, \ t \in [0,T], \tag{2.7}$$

where the velocity parameter is given as $v = (v_1, \ldots, v_d)^T \in \mathbb{R}^{d,+}$, the reaction parameter is given as $\lambda \in \mathbb{R}^+$ and the spatial variables are $x = (x_1, \ldots, x_d)^T \in \Omega \subset \mathbb{R}^{d,+}$.

After space and time discretizations with first-order finite difference methods in time and space, we can derive the following CFL conditions for $d = 2$:

$$CFL_{\text{flow},x_1} = |\tfrac{v_{x_1}\,\tau}{\Delta x_1}| \leq 1,$$

$$CFL_{\text{flow},x_2} = |\tfrac{v_{x_2}\,\tau}{\Delta x_2}| \leq 1,$$

$$CFL_{\text{react}} = |\lambda\,\tau| \leq 1,$$

where τ is the time step, Δx_1 and Δx_2 are the spatial steps.

Here we have the relation between the spatial and the time step of each operator are given as:

$$\tau_{x_1} \quad \leq \tfrac{\Delta x_1}{v_{x_1}}, \tag{2.8}$$

$$\tau_{x_2} \quad \leq \tfrac{\Delta x_2}{v_{x_2}}, \tag{2.9}$$

$$\tau_\lambda \quad \leq \tfrac{1}{\lambda}, \tag{2.10}$$

where τ_{x_1}, τ_{x_2}, and τ_λ are the time steps of the flow operators and reaction operator.

If we assume for example the following timescales:

$$\tau_{x_1} \approx \tau_{react} >> \tau_{x_2}. \tag{2.11}$$

We solve the operator A, which consists of the x_1-direction of the flow and the reaction term, with the time step τ_{x_1}. Further the operator B, which consists of the x_2-direction of the flow, is solved with the time step τ_{x_2}, see also the splitting schemes in [153] and [183].

Remark 2.1. Due to the discretization methods and their underlying stability criteria, there are further conditions for special evolution equations, such as the Neumann number for the diffusion equation and the Prandtl or Reynolds number to characterize the Navier-Stokes equation. These conditions present the physical behavior in terms of stability and can be used as a decoupling criterion, see [150], [183].

2.2.2 Mathematical Decomposition

In the mathematical decomposition, we consider ordinary differential equation, which are given directly, or spatially discretized partial differential equations, e.g., discretized by methods of lines, see [117].

Here, we assume that we decouple with respect to the spatial dimensions the following convection-diffusion equation in two dimensions:

$$\frac{\partial u}{\partial t} = A_x u + A_y u, \quad (x, y) \in \Omega, \, t \in [0, T], \tag{2.12}$$

$$u(x, y, 0) = u_0(x, y), \quad (x, y) \in \Omega,$$

$$u(x, y, t) = g(x, y, t), \quad (x, y) \in \partial\Omega, \, t \in [0, T],$$

$$A_x u = -v_x \frac{\partial u}{\partial x} + \frac{\partial}{\partial x} D_x \frac{\partial u}{\partial x}, \tag{2.13}$$

$$A_y u = -v_y \frac{\partial u}{\partial y} + \frac{\partial}{\partial y} D_y \frac{\partial u}{\partial y}, \tag{2.14}$$

where A_x is the operator in the x-direction and A_y is the operator in the y-direction. The parameters are given as $v_x, v_y, D_x, D_y \in \mathbb{R}^+$.

Here the underlying idea is to decouple into different spatial directions, where each direction can be discretized separately into a one-dimensional method.

Remark 2.2. The operator splitting method can be seen as a locally one-dimensional method, e.g., see LOD (locally one dimensional) [40], [45], and [167]. We assume we have one direction, e.g., the x-direction, which is stiff and the y-direction, which is not stiff. For such a relation, the separation into two different schemes makes sense. Based on the separation, we can approximate each direction with the optimal timestep, that is given with the CFL condition, see Equations (2.8)–(2.10). Such ideas are implied to higher-order implicit-explicit Runge-Kutta schemes, while small timescale operators are solved with the implicit part and large timescale operators are solved with the explicit part, see also [91] and [192].

2.3   Introduction to Classical Splitting Methods

The natural way of decoupling a linear ordinary differential equation with constant coefficients into simpler parts is the following way,

$$\frac{du(t)}{dt} = A_{\text{full}} \, u(t) \, , \quad t \in (0, T) \, , \tag{2.15}$$

$$\frac{du(t)}{dt} = (A + B) \, u(t) \, , \quad t \in (0, T) \, , \tag{2.16}$$

where $A_{\text{full}}, A, B : \mathbb{R}^m \to \mathbb{R}^m$ are matrices, $u = (u_1, \ldots, u_m)^T$ is the solution vector, and m is a given positive number. The initial conditions are given as $u(t = 0) = u_0$, while u_0 is a given constant vector. The operator A_{full} can be decoupled into the operators A and B, cf. the introduction in [192].

Based on these linear operators (which are matrices), equation (2.15) can be solved exactly. The solution is given as

$$u(t^{n+1}) = \exp(\tau A_{\text{full}}) \, u(t^n) \,, \tag{2.17}$$

where the time step is $\tau = t^{n+1} - t^n$.

The simplest splitting methods are the sequential splitting methods that are decoupled into two or more equations.

For linear and bounded operators, we can analyze the splitting error by performing a Taylor expansion, see [20].

Remark 2.3. This introduction to ordinary differential equations can be extended to the abstract Cauchy problem of a parabolic equation and the possibility of defining an operator A_{full} as a Friedrichs extension, see [2] and [11]. Therefore, weak solutions are possible and we can also apply the notation to exp-formulations, see [11].

We now deal with the following classical formulation of splitting methods.

2.3.1 Classical Formulation of Splitting Methods

In the following, we describe the traditional splitting methods that are widely used for the solution of real-life problems. We focus our attention on the case of two linear operators, i.e., we consider the following Cauchy problem

$$\frac{du(t)}{dt} = Au(t) + Bu(t), \quad t \in (0, T), \quad u(0) = u_0, \tag{2.18}$$

whereby the initial function u_0 is given, and A and B are assumed to be bounded linear operators in the Banach space $\mathbf{X}$ with $A, B : \mathbf{X} \to \mathbf{X}$. We also have a norm $|| \cdot ||_{\mathbf{X}}$ corresponding to the space $\mathbf{X}$, e.g., the Euclidean norm if A and B are matrices. In realistic applications, the operators are unbounded and correspond to physical operators, e.g., convection or diffusion operator, the proofs based on the Taylor expansion are no longer possible and we have to change the techniques of the proof to analytic semigroups, see [120], [121], and [136].

2.3.2 Sequential Splitting Method

First, we describe the simplest operator splitting, called the *sequential splitting method*. The sequential splitting method solves two subproblems sequentially on subintervals $[t^n, t^{n+1}]$, where $n = 0, 1, \ldots, N - 1$, $t^0 = 0$ and $t^N = T$. The different subproblems are connected by the initial conditions. This means that we replace the original problem (2.18) with the subproblems at subintervals

$$\frac{\partial u^*(t)}{\partial t} = Au^*(t), \quad t \in (t^n, t^{n+1}), \quad \text{with } u^*(t^n) = u_{\text{sp}}^n, \tag{2.19}$$

$$\frac{\partial u^{**}(t)}{\partial t} = Bu^{**}(t), \quad t \in (t^n, t^{n+1}), \quad \text{with } u^{**}(t^n) = u^*(t^{n+1}), \tag{2.20}$$

for $n = 0, 1, \ldots, N-1$, where $u_{\mathrm{sp}}^0 = u_0$ is given from (2.18). The approximate split solution at the point $t = t^{n+1}$ is defined as $u_{\mathrm{sp}}^{n+1} = u^{**}(t^{n+1})$.

Clearly, converting the original problems to subproblems usually results in some error, called the *local splitting error*. The local splitting error of the sequential operator splitting method can be derived as

$$
\begin{aligned}
\mathrm{err}_{\mathrm{local}}(\tau_n) &= (\exp(\tau_n(A+B)) - \exp(\tau_n B)\exp(\tau_n A))\, u_{\mathrm{sp}}^n \\
&= \frac{1}{2}\tau_n^2 \|[A, B]\, u_{\mathrm{sp}}^n\| + \mathcal{O}(\tau_n^3), &&(2.21)
\end{aligned}
$$

$$
\begin{aligned}
\mathrm{err}_{\mathrm{global}}(\tau_n) &= (n+1)\tau_n \frac{\mathrm{err}_{\mathrm{local}}(\tau_n)}{\tau_n} \\
&= \frac{1}{2}t^{n+1}\tau_n \|[A, B]\, u_{\mathrm{sp}}^n\| + \mathcal{O}(\tau_n^2), &&(2.22)
\end{aligned}
$$

where the operators A and B are assumed to be bounded operators. The splitting time step is defined as $\tau_n = t^{n+1} - t^n$. We define $[A, B] := AB - BA$ as the commutator of A and B. Consequently, the splitting error is $\mathcal{O}(\tau_n)$ (first order in time), when the operators A and B do not commute. When the operators commute, then the method is exact. Hence, by definition, sequential operator splitting is called a *first-order splitting method*.

Remark 2.4. A characteristic of traditional splitting methods is the relation between the commutator and the consistency of the method, see references [20], and [56].

We assume the simple sequential splitting method given in (2.19) and (2.20), where A and B are bounded operators in a Banach space $\mathbf{X}$.

So we obtain the local splitting error of a simple sequential splitting method with respect to the commutator formulation:

$$
\begin{aligned}
\|\mathrm{err}_{\mathrm{local}}(\tau_n)\| &= \|(\exp(\tau_n(A+B)) - \exp(\tau_n B)\exp(\tau_n A))\, u_{\mathrm{sp}}^n\| \\
&= \frac{1}{2}\|[A, B]\, u_{\mathrm{sp}}^n\|\tau_n^2 + \mathcal{O}(\tau_n^3), &&(2.23)
\end{aligned}
$$

where $[A, B] \neq \mathbf{0}$ ($\mathbf{0}$ is the zero matrix) if the operators do not commute.

Based on the commutator $[A, B]$, we have a large constant in the local splitting error of $\mathcal{O}(\tau_n^2)$ and therefore we have to use unrealistically small time steps, [82].

The same can also be derived for higher-order classical splitting methods, as presented in the next section.

Remark 2.5. The main issue with splitting problems is the reduction of order for applications with stiff and nonstiff operators. We reduce one order if one operator is stiff compared to the other operator. So we reduce the consistency error; such problems are known and discussed in [151], [152], and [198].

Example 2.6. *We consider a constant linear system of ODEs:*

$$\frac{du}{dt} = Au + Bu, \ t \in [0, T], \tag{2.24}$$
$$u(0) = u_0,$$

where A is supposed to be stiff and B nonstiff.

By stiffness, we mean that τA has a huge norm over the range of step sizes τ_n, see also [39] and [123].

We assume:

$$\|\tau_n A\| \gg 1, \ \|\tau_n B\| = \mathcal{O}(\tau_n).$$

This means that the stiff operator A has large eigenvalues and requires smaller time steps to obtain accurate results, while the operator B has small eigenvalues and can accommodate larger time steps to the same accuracy, see [152, 198].

Such problems require at least a higher-order time discretization method, e.g., Runge-Kutta (RK) method of higher order.

2.3.3   Symmetrical Weighted Sequential Splitting Methods

For noncommuting operators, sequential operator splitting is not symmetric with respect to the operators A and B and it has first-order accuracy. However, in many practical cases, we require splittings of higher-order accuracy. We can achieve this with the following modified splitting method, called the *symmetrical weighted sequential splitting method*, which is already symmetrical with respect to the operators.

For the algorithm, we consider again the Cauchy problem (2.18) and define the splitting of the operator on the time interval $[t^n, t^{n+1}]$ (where $t^{n+1} = t^n + \tau_n$) as

$$\frac{\partial u^*(t)}{\partial t} = Au^*(t), \ t \in (t^n, t^{n+1}), \quad \text{with } u^*(t^n) = u_{\text{sp}}^n, \tag{2.25}$$

$$\frac{\partial u^{**}(t)}{\partial t} = Bu^{**}(t), \ t \in (t^n, t^{n+1}), \quad \text{with } u^{**}(t^n) = u^*(t^{n+1}),$$

and

$$\frac{\partial v^*(t)}{\partial t} = Bv^*(t), \ t \in (t^n, t^{n+1}), \quad \text{with } v^*(t^n) = u_{\text{sp}}^n, \tag{2.26}$$

$$\frac{\partial v^{**}(t)}{\partial t} = Av^{**}(t), \ t \in (t^n, t^{n+1}), \quad \text{with } v^{**}(t^n) = v^*(t^{n+1}),$$

where u_{sp}^n is known.

The approximation at the next time level t^{n+1} is defined as

$$u_{\text{sp}}^{n+1} = \frac{u^{**}(t^{n+1}) + v^{**}(t^{n+1})}{2}. \tag{2.27}$$

The splitting error of this splitting method is derived as follows:

$$\text{err}_{\text{local}}(\tau_n) = \left|\left|\left(\exp(\tau_n(A+B)) - \right.\right.\right. \tag{2.28}$$

$$\left.\left.\left. -\frac{1}{2}[\exp(\tau_n B)\exp(\tau_n A) + \exp(\tau_n A)\exp(\tau_n B)]\right)c_{\text{sp}}^n\right|\right|,$$

where $||\cdot||$ is a Banach norm (e.g., here the Euclidean norm for A, B are matrices).

An easy computation shows that in the general case of bounded operators we have

$$\text{err}_{\text{global}}(\tau_n) = \mathcal{O}(\tau_n^2), \tag{2.29}$$

i.e., the method is of second-order accuracy. We note that in the case of commuting operators A and B the method is exact, i.e., the splitting error vanishes.

Remark 2.7. For unbounded operators, the proof is shown in [56]. The symmetrically weighted sequential splitting method is an additive splitting method and can be locally decoupled into two separate equations. This method is simple to parallelize; see for example applications in large-scale computation, e.g., [204].

2.3.4 Strang-Marchuk Splitting Method

One of the most popular and widely used operator splittings is the so-called *Strang splitting (or Strang-Marchuk splitting)*, see [194], which takes the following form

$$\frac{\partial u^*(t)}{\partial t} = Au^*(t), \text{ with } t^n \leq t \leq t^{n+1/2} \text{ and } u^*(t^n) = u_{\text{sp}}^n, \tag{2.30}$$

$$\frac{\partial u^{**}(t)}{\partial t} = Bu^{**}(t), \text{ with } t^n \leq t \leq t^{n+1} \text{ and } u^{**}(t^n) = u^*(t^{n+1/2}),$$

$$\frac{\partial u^{***}(t)}{\partial t} = Au^{***}(t), \text{ with } t^{n+1/2} \leq t \leq t^{n+1} \text{ and } u^{***}(t^{n+1/2}) = u^{**}(t^{n+1}),$$

where $t^{n+1/2} = t^n + \frac{1}{2}\tau_n$, and the approximation on the next time level t^{n+1} is defined as $u_{\text{sp}}^{n+1} = u^{***}(t^{n+1})$.

The splitting error of Strang splitting is

$$\text{err}_{\text{local}}(\tau_n) = \frac{1}{24}\tau_n^3||([B,[B,A]] - 2[A,[A,B]]) u_{\text{sp}}^n|| + \mathcal{O}(\tau_n^4), \tag{2.31}$$

$$\text{err}_{\text{global}}(\tau_n) = \frac{1}{24} t^{n+1} \tau_n^2||([B,[B,A]] - 2[A,[A,B]]) u_{\text{sp}}^n||$$

$$+ \mathcal{O}(\tau_n^3), \tag{2.32}$$

see e.g., [194]. We see that this operator splitting is also of second order. We note that under some special conditions for the operators A and B, the Strang splitting has third-order accuracy and can even be exact, see [194].

In the next section, we present some other types of operator splitting methods, that are based on a combination of operator splitting and iterative methods.

2.3.5   Higher-Order Splitting Methods

The higher-order splitting methods are used for more accurate computations, but also to have more computational steps. These methods are often performed in quantum dynamics to approximate the evolution operator $\exp(\tau_n(A + B))$, see [29].

An analytical construction of higher-order splitting methods can be performed with the help of the Baker-Campbell-Hausdorff (BCH) formula, see [124], [202].

The reconstruction process is based on the following product of exponential functions,

$$
\begin{aligned}
\exp(\tau_n(A + B)) &= \Pi_{i=1}^{m} \exp(c_i \tau_n A) \exp(d_i \tau_n B) + \mathcal{O}(\tau_n^{m+1}) \quad (2.33) \\
&= S(\tau_n),
\end{aligned}
$$

where A, B are noncommutative operators, τ_n is an equidistant time step, and $(c_1, c_2, \ldots)$, $(d_1, d_2, \ldots)$ are real numbers. $S(\tau_n)$ is a group that is generated by the operator $(A + B)$. The product of exponential functions is called the *integrator*.

Thus, for the construction of a first-order method, we have the trivial solution:

$c_1 = d_1 = 1$ and $m = 1$.

For a fourth-order method, see [171], we have the following coefficients:

$$
c_1 = c_4 = \frac{1}{2(2 - 2^{1/3})} \ , \ c_2 = c_3 = \frac{1 - 2^{1/3}}{2(2 - 2^{1/3})},
$$

$$
d_1 = d_3 = \frac{1}{2 - 2^{1/3}}, \ d_2 = -\frac{2^{1/3}}{2 - 2^{1/3}}, \ d_4 = 0.
$$

We can improve this direct method by using symmetric integrators and obtain orders of 4, 6, 8, ..., see [202]. For this construction, exact reversibility in time is important, i.e. $S(\tau_n)S(-\tau_n) = S(-\tau_n)S(\tau_n)$.

Remark 2.8. The construction of higher-order methods is based on forward and backward time steps, due to the assumption of time reversibility. *Negative time steps are also possible.* A visualization of the splitting steps is presented in Figure 2.1.

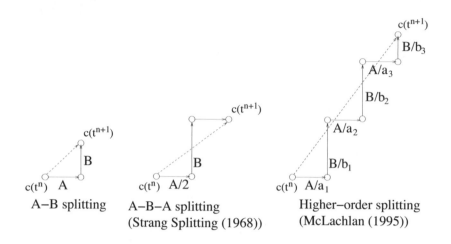

FIGURE 2.1: Graphical visualization of noniterative splitting methods.

Remark 2.9. The construction of higher-order methods can also be made with symplectic operators, e.g., conservation of the symplectic two-form $dp \wedge dq$, see [202]. Based on this symplectic behavior, the total energy is guaranteed.

Remark 2.10. Physical restrictions, such as the status of temporal irreversibility, e.g., the quantum statistical trace or imaginary temporal Schröder equation, will also require positive coefficients in higher-order methods. Based on these restrictions, backward time steps are not allowed. Theorems such as special higher-order methods have been developed, see [28], [30], [97], and [105]. The general idea is to force some coefficients, and thus their underlying commutators, to be zero.

Remark 2.11. The main problem of such a single product splitting method is that beyond second-order accuracy, one requires an exponentially growing number of operators with unavoidable *negative* coefficients, see [191, 195]. Such methods cannot be applied to time-irreversible or semigroup problems. Even for time-reversible systems where negative time steps do not pose a problem, the exponential growth in the number of force evaluations renders high-order symplectic integrators difficult to derive and expensive to use. Such drawbacks can be overcome with iterative splitting methods, which are discussed extensively in the next section.

2.4 Iterative Splitting Method

Iterative splitting method underlie the iterative methods used to solve coupled operators by using a fixed-point iteration. These algorithms integrate each underlying equation with respect to the last iterated solution. Therefore, the starting solution in each iterative equation is important in order to guarantee fast convergence or a higher-order method. The last iterative solution should at least have a local error of $\mathcal{O}(\tau_n^i)$ (i-th order in time), where i is the number of iteration steps, to obtain the next higher order.

We deal with at least two equations, therefore two operators, but the results can be generalized to n operators (see for example the ideas of waveform-relaxation methods [199]).

In our next analysis, we deal with the following problem:

$$
\begin{aligned}
\frac{du(t)}{dt} &= Au(t) + Bu(t), \text{ for } 0 \leq t \leq T, \qquad (2.34)\\
u(0) &= u_0,
\end{aligned}
$$

where A, B are bounded linear operators. For such a problem, we can derive the analytical solution, given as

$$
u(t) = \exp((A + B)t)\, u_0, \text{ for } 0 \leq t \leq T. \qquad (2.35)
$$

We propose the iterative splitting method as a decomposition method, which is an effective solver for large systems of partial differential equations.

The iterative splitting method belongs to a second group of iterative methods for solving coupled equations. We can combine the traditional splitting method (decoupling the time interval into smaller parts according to the splitting time step) with the iterative splitting method (on each split time interval, we use one-step iterative methods). At the very least, iterative splitting method serves as a predictor-corrector method, so in the first equation the solution is predicted, whereas in the second equation the solution is corrected, see [144].

We use the iterative splitting method, while standard splitting methods have, in addition to its benefits, several drawbacks:

- For noncommuting operators, there may be a very large constant local splitting error, requiring the use of an unrealistically small splitting time step. In other words, the stability and commutativity are connected by the norm of the commutator, see Remark 2.13.

- Within a full splitting step in one subinterval, the inner values are not an approximation to the solution of the original problem.

- Splitting the original problem into different subproblems with one operator (i.e., neglecting the other components) is physically correct, see for example Strang splitting. But the method is physically questionable when we aim to get consistent approximations after each inner step, because we lose the exact starting conditions.

- Beyond second-order accuracy, an exponentially growing number of operators with negative coefficients are necessary. To obtain higher-order splitting methods is a delicate problem, examples are complex coefficients, see [120, 121], or multiproduct expansion methods (extrapolation methods), see [30] and [97].

- Efficient and fast Multistep methods are delicate to implement for such splitting schemes, while subintervals are not an approximation of the original problem.

Thus, for the iterative splitting method, we state the following theses.

- For noncommuting operators, we may reduce the local splitting error, by using more iteration steps to obtain higher-order accuracy.

- We must solve the original problem within a full splitting step, while keeping all operators in the equations.

- Splitting the original problem into different subproblems, including all operators of the problem, is physically the best. We obtain consistent approximations after each inner step, because of the exact or approximate starting conditions of the previous solution in the iteration.

- Higher-order schemes can be achieved with additional iteration steps.

- Depending on the splitting scheme, we have to deal with explicit or implicit operators. The first type of operator is cheap to compute (explicit), while the other operator can be computed with a known implicit method.

- Efficient and fast Multistep methods, e.g., BDF (Backward Differentiation Formula) or SBDF (Stiff Backward Differentiation Formula) methods, are simple to implement and accelerate the solver process, see [98].

Remark 2.12. One of the main differences between the standard splitting methods, is in the solution of an inhomogeneous differential equation as opposed to a homogeneous differential equation. So a certain amount of computation is involved in dealing with the inhomogeneous part using the explicit operator given. Here, the idea of overcoming such a problem is to use fast explicit solvers that are not time-intensive at solving the inhomogeneous part, see an application to the reaction parts that can be solved fast [106].

Remark 2.13. The commutator is related to the consistency of the method. We assume a sequential splitting method:

$$\frac{du_1(t)}{dt} = Au_1(t), \; u_1(t^n) = u^n, \tag{2.36}$$

$$\frac{du_2(t)}{dt} = Bu_2(t), \; u_2(t^n) = u_1(t^{n+1}), \tag{2.37}$$

where A, B are bounded operators in a Banach space $\mathbf{X}$ with Banach norm $|| \cdot ||_{\mathbf{X}} = || \cdot ||$. The time step is $\tau = t^{n+1} - t^n$, with $t \in [t^n, t^{n+1}]$. The result for the splitting method is given as $u(t^{n+1}) \approx u_2(t^{n+1}) = u_{\mathrm{sp}}(t^{n+1})$.

Then, we obtain the local error:

$$||\mathrm{err}_{\mathrm{local}}(\tau)|| = ||(\exp((A + B)_t) - \exp(A\tau)\exp(B\tau))u_0||, \tag{2.38}$$

$$\leq |||[A, B]||| \, \mathcal{O}(\tau^2) \, ||u_0||, \tag{2.39}$$

where $\mathrm{err}_{\mathrm{local}}(\tau) = u(t^{n+1}) - u_{\mathrm{sp}}(t^{n+1})$ is defined and, for stability, the commutator in the matrix norm $|| \cdot ||_{\mathbf{X}} = || \cdot ||$, e.g., the maximum norm, must be bounded, e.g., $|||[A, B]||| < C$, where $C \in \mathbb{R}^+$ is a constant.

2.4.0.1 Iterative Splitting Method (Algorithm)

In order to avoid the problems mentioned above, we can use iterative splitting on the interval $[0, T]$, cf. [144]. In the following discussion, we suggest a modification of this method by introducing a splitting time discretization. We suggest an algorithm based on the iteration of a fixed sequential splitting discretization with step size τ_n. On the time interval $[t^n, t^{n+1}]$, we solve the following subproblems consecutively for $i = 1, 3, 5, \ldots 2m + 1$.

$$\frac{du_i(t)}{dt} = Au_i(t) + Bu_{i-1}(t), \tag{2.40}$$

$$t \in (t^n, t^{n+1}), \quad \text{with } u_i(t^n) = u_{\mathrm{sp}}^n,$$

$$\frac{du_{i+1}(t)}{dt} = Au_i(t) + Bu_{i+1}(t), \tag{2.41}$$

$$t \in (t^n, t^{n+1}), \quad \text{with } u_{i+1}(t^n) = u_{\mathrm{sp}}^n,$$

where $u_0(t)$ is any fixed function for each iteration. The initial solution can be given as:

- $u_0(t) = 0$ (we initialize with zero),

- $u_0(t) = u(t^n)$ (we initialize with the old solution at time t^n),

where we achieve one order more with the second version and can accelerate the solver process, see [55] and [99]. While the initial condition to the next iteration process is important and reflected in the error term, we have taken into account more accurate initial conditions. For a more accurate initialization process, we can modify the iterative scheme to a weighted iterative scheme, which is discussed in [81] and [75].

(Here, as before, u_{sp}^n denotes the known split approximation at time level $t = t^n$). The split approximation at time level $t = t^{n+1}$ is defined as $u_{\text{sp}}^{n+1} = u_{2m+2}(t^{n+1})$.

The algorithm (2.40)–(2.41) is an iterative method, which at each step consists of both operators A and B. Hence, in these equations, there is no real separation of the different physical processes. However, we note that subdividing the time interval into subintervals distinguishes this process from the simple fixed-point iteration and turns it into a more efficient numerical method.

We also observe that the algorithm (2.40)–(2.41) is a real operator splitting method, because Equation (2.40) requires a problem with operator A solved, and (2.41) requires a problem with operator B solved. Hence, as in sequential operator splitting, the two operators are separated.

Assumption 2.14. *In the next subsections, we assume that the problem involves solving differential Equations (2.40) and (2.41) exactly or at least with the same or higher order as the underlying order of the splitting method. Based on this assumption, we can concentrate on the error of the splitting method, while the error of the underlying numerical method used to solve the differential equation is less than the error of the splitting method under consideration, see [144].*

2.5   Consistency Analysis of Iterative Splitting Method

In this subsection, we analyze the consistency and order of the iterative splitting method. First, in Section 2.5.1 we consider the original algorithm (2.40)–(2.41), prove its consistency, and define the order of the local splitting error.

The algorithm (2.40)–(2.41) requires a knowledge of the functions $c_{i-1}(t)$ and $c_i(t)$ over the whole interval $[t^n, t^{n+1}]$, which is typically not known, because generally their values are only known at several points of the split interval. Hence, typically we can only define interpolations of these functions. In Section 2.5.2, we prove the consistency of such a modified algorithm.

2.5.1   Local Error Analysis

Here we will analyze the consistency and order of the local splitting error of method (2.40)–(2.41) for linear bounded operators $A, B : \mathbf{X} \to \mathbf{X}$, where $\mathbf{X}$

is a Banach space. In the following, we use the notation $\mathbf{X}^2$ for the product space $\mathbf{X} \times \mathbf{X}$ supplied with the norm $\|(u,v)^T\|_\infty = \max\{\|u\|, \|v\|\}$ $(u,v \in \mathbf{X})$.

We have the following order of consistency for our iterative splitting method.

Theorem 2.15. *Let $A, B \in \mathcal{L}(\mathbf{X})$ be given linear bounded operators. We consider the abstract Cauchy problem:*

$$\partial_t u(t) = Au(t) + Bu(t), \quad 0 < t \le T, \tag{2.42}$$
$$u(0) = u_0. \tag{2.43}$$

Then problem (2.42) has a unique solution. The iteration (2.40)–(2.41) for $i = 1, 3, \ldots, 2m+1$ is consistent with an order of consistency $\mathcal{O}(\tau_n^{2m+1})$.

Proof. Because $A + B \in \mathcal{L}(\mathbf{X})$, therefore it is a generator of a uniformly continuous semigroup; hence, problem (2.42) has a unique solution $u(t) = \exp((A+B)t)u_0$.

Let us consider the iteration (2.40)–(2.41) on the subinterval $[t^n, t^{n+1}]$. For the local error function $e_i(t) = u(t) - u_i(t)$, we have the following relations:

$$\partial_t e_i(t) = Ae_i(t) + Be_{i-1}(t), \quad t \in (t^n, t^{n+1}], \tag{2.44}$$
$$e_i(t^n) = 0,$$

and

$$\partial_t e_{i+1}(t) = Ae_i(t) + Be_{i+1}(t), \quad t \in (t^n, t^{n+1}], \tag{2.45}$$
$$e_{i+1}(t^n) = 0,$$

for $i = 1, 3, 5, \ldots$, with $e_1(0) = 0$ and $e_0(t) = u(t)$. The elements $\mathcal{E}_i(t)$, $\mathcal{F}_i(t) \in \mathbf{X}^2$ and the linear operator $\mathcal{A} : \mathbf{X}^2 \to \mathbf{X}^2$ are defined as follows:

$$\mathcal{E}_i(t) = \begin{bmatrix} e_i(t) \\ e_{i+1}(t) \end{bmatrix}, \quad \mathcal{F}_i(t) = \begin{bmatrix} Be_{i-1}(t) \\ 0 \end{bmatrix}, \quad \mathcal{A} = \begin{bmatrix} A & 0 \\ A & B \end{bmatrix} \tag{2.46}$$

Then, using the notations (2.46), the relations (2.44)–(2.45) can be written in the form

$$\partial_t \mathcal{E}_i(t) = \mathcal{A}\mathcal{E}_i(t) + \mathcal{F}_i(t), \quad t \in (t^n, t^{n+1}], \tag{2.47}$$
$$\mathcal{E}_i(t^n) = 0.$$

Because of our assumptions, $\mathcal{A}$ is a generator of the one-parameter C_0 semigroup $(\exp \mathcal{A}t)_{t \ge 0}$. Hence, by using variations of the constants formula, the solution of the abstract Cauchy problem (2.47) with homogeneous initial conditions can be written as:

$$\mathcal{E}_i(t) = \int_{t^n}^t \exp(\mathcal{A}(t-s))\mathcal{F}_i(s)ds, \quad t \in [t^n, t^{n+1}]. \tag{2.48}$$

Then, using the notation

$$\|\mathcal{E}_i\|_\infty = \sup_{t \in [t^n, t^{n+1}]} \|\mathcal{E}_i(t)\|, \tag{2.49}$$

we have

$$\|\mathcal{E}_i(t)\| \;\leq\; \|\mathcal{F}_i\|_\infty \int_{t^n}^t \|\exp(\mathcal{A}(t-s))\| ds \tag{2.50}$$

$$= \; \|B\| \|e_{i-1}\| \int_{t^n}^t \|\exp(\mathcal{A}(t-s))\| ds, \quad t \in [t^n, t^{n+1}].$$

Because $(\mathcal{A}(t))_{t \geq 0}$ is a semigroup, therefore the so-called *growth estimation*,

$$\|\exp(\mathcal{A}t)\| \leq K \exp(\omega t); \quad t \geq 0, \tag{2.51}$$

holds with some numbers $K \geq 0$ and $\omega \in \mathbb{R}$.

- Assume that $(\mathcal{A}(t))_{t \geq 0}$ is a bounded or exponentially stable semigroup, i.e., (2.51) holds for some $\omega \leq 0$. Then, obviously, the estimate

$$\|\exp(\mathcal{A}t)\| \leq K, \quad t \geq 0, \tag{2.52}$$

  holds, and hence, according to (2.50), we have the relation

$$\|\mathcal{E}_i\|(t) \leq K \|B\| \tau_n \|e_{i-1}\|, \quad t \in [t^n, t^{n+1}]. \tag{2.53}$$

- Assume that $(\exp \mathcal{A}t)_{t \geq 0}$ has an exponential growth with some $\omega > 0$. Using (2.51) we have

$$\int_{t^n}^t \|\exp(\mathcal{A}(t-s))\| ds \leq K_\omega(t), \quad t \in [t^n, t^{n+1}], \tag{2.54}$$

  where

$$K_\omega(t) = \frac{K}{\omega} \left(\exp(\omega(t-t^n)) - 1 \right), \quad t \in [t^n, t^{n+1}]. \tag{2.55}$$

  Hence

$$K_\omega(t) \leq \frac{K}{\omega} \left(\exp(\omega \tau_n) - 1 \right) = K \tau_n + \mathcal{O}(\tau_n^2). \tag{2.56}$$

The estimations (2.53) and (5.61) result in

$$\|\mathcal{E}_i\|_\infty \leq K \|B\| \tau_n \|e_{i-1}\| + \mathcal{O}(\tau_n^2). \tag{2.57}$$

Taking into account the definition of $\mathcal{E}_i$ and the maximum norm $\| \cdot \|_\infty$, we obtain

$$\|e_i\| \leq K \|B\| \tau_n \|e_{i-1}\| + \mathcal{O}(\tau_n^2), \tag{2.58}$$

and consequently,

$$\begin{aligned}
\|e_{i+1}\| &\leq K\|B\|\|e_i\| \int_{t^n}^{t} \|\exp(\mathcal{A}(t-s))\| ds, &(2.59)\\
&\leq K\|B\| \, \tau_n \, (K\|B\|\tau_n\|e_{i-1}\| + \mathcal{O}(\tau_n^2)),\\
&\leq K_2\tau_n^2\|e_{i-1}\| + \mathcal{O}(\tau_n^3).
\end{aligned}$$

We apply a recursive argument that proves our statement. □

Remark 2.16. When A and B are matrices (i.e., when (2.40)–(2.41) is a system of ordinary differential equations), we can use the concept of logarithmic norm for estimation of growth (2.51). Hence, for many important classes of matrices, we can prove the validity of (2.51) with $\omega \leq 0$.

Remark 2.17. We note that a huge class of important differential operators generate a contractive semigroup. This means that for such problems, assuming the exact solvability of the split subproblems, the iterative splitting method converges on the exact solution to the second order.

Remark 2.18. We note that the assumption $A \in \mathcal{L}(X)$ can be formulated more weakly as it is enough to assume that the operator A is the generator of a C_0 semigroup.

Remark 2.19. When T is a sufficiently small number, then we do not need to partition the interval $[0, T]$ into subintervals. In this case, the convergence of the iteration (2.40)–(2.41) to the solution of the problem (2.42) follows immediately from Theorem 2.15, and the rate of convergence is equal to the order of the local splitting error.

Remark 2.20. Estimate (2.60) shows that, after the final iteration step ($i = 2m + 1$), we have the estimation

$$\|e_{2m+1}\| \leq K_m\|e_0\|\tau_n^{2m} + \mathcal{O}(\tau_n^{2m+1}). \qquad (2.60)$$

This relation shows that the constant in the leading term strongly depends on the choice of initial guess $u_0(t)$. When the choice is $u_0(t) = 0$ (see [144]), then $\|e_0\| = u(t)$ (where $u(t)$ is the exact solution of the original problem) and hence the error might be very significant. Such a problem can be circumvented, if we start with an improved solution. For example, we achieve one order more, when the choice of the initial guess is $u_0(t) = u(t^n)$. This means we start with the previous solution on time t^n, see [55]. Higher-order improvements can be done with additional prestepping schemes, which are approximations to the exact solution, see [98].

Remark 2.21. In realistic applications, the final iteration steps $2m + 1$ and the time step τ_n are chosen in optimal relation to one another, such that the time step τ_n is chosen to be maximal and with at least 3 or 5 iteration steps, see Section 2.5.1. In addition, a final stop criterion as an error bound, e.g.

$|u_i - u_{i-1}| \leq$ err with for example err $= 10^{-4}$, helps to restrict the number of steps. A graphical illustration of the iterative splitting method is given in Figure 2.2.

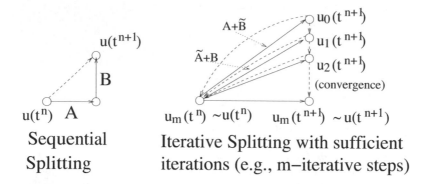

FIGURE 2.2: Graphical visualization of noniterative and iterative splitting methods.

2.5.2   Increasing the Order of Accuracy with Improved Initial Functions and Consistency Analysis

We can increase the order of accuracy by improving our choice of initial iteration function, see [54].

Based on our previous assumption about the initial solutions, we start with an exact solution or interpolated split solution and present our theory for the exactness of the method.

Exact Solution of the Split Subproblem

We derive the exact solution of Equations (2.40) and (2.41) by solving the first split problem,

$$u_i(t^{n+1}) = \exp(At)u^n + \sum_{s=0}^{\infty} \sum_{k=s+1}^{\infty} \frac{t^k}{k!} A^{k-s-1} B u_{i-1}^{(s)}(t^n), \qquad (2.61)$$

and the second split problem,

$$u_{i+1}(t^{n+1}) = \exp(Bt)u^n + \sum_{s=0}^{\infty} \sum_{k=s+1}^{\infty} \frac{t^k}{k!} B^{k-s-1} A u_i^{(s)}(t^n), \qquad (2.62)$$

where $\tau_n = t^{n+1} - t^n$ is the equidistant time step and $u^n = u(t^n)$ is the exact solution at time t^n or at least approximately of local order $\mathcal{O}(\tau_n^{m+2})$. n is the number of time steps ($n \in \{0, \ldots, N\}, N \in \mathbb{N}^+$) and $m > 0$ is the number of iteration steps.

Theorem 2.22. *Assume that for functions $u_{i-1}(t^{n+1})$ and $u_i(t^{n+1})$ the conditions*

$$u_{i-1}^s(t^n) = (A+B)^s u^n, \ s = 0, 1, \ldots, m+1, \tag{2.63}$$

$$u_i^s(t^n) = (A+B)^s u^n, \ s = 0, 1, \ldots, m+2, \tag{2.64}$$

are satisfied. After $m+2$ iterations, the method has a local splitting error $\mathcal{O}(\tau_n^{m+2})$ and therefore the global error $\mathrm{err}_{\mathrm{global}}$ is $\mathcal{O}(\tau_n^{m+1})$.

Proof. We show that

$$\exp(\tau_n(A+B))u^n - u_{m+1}(t^{n+1}) = \mathcal{O}(\tau_n^{m+1}), \tag{2.65}$$

$$\exp(\tau_n(A+B))u^n - u_{m+2}(t^{n+1}) = \mathcal{O}(\tau_n^{m+2}). \tag{2.66}$$

Based on the assumption and using the exact solutions (2.61) and (2.62), we must prove the relations:

$$\sum_{p=0}^{m+1} \frac{1}{p!}\tau_n^p(A+B)^p \tag{2.67}$$

$$= \sum_{p=0}^{m+1} \frac{1}{p!}\tau_n^p(A)^p + \sum_{s=0}^{m}\sum_{k=s+1}^{m+1} \frac{\tau_n^k}{k!}A^{k-s-1}B,$$

and

$$\sum_{p=0}^{m+2} \frac{1}{p!}\tau_n^p(A+B)^p \tag{2.68}$$

$$= \sum_{p=0}^{m+2} \frac{1}{p!}\tau_n^p(B)^p + \sum_{s=0}^{m+1}\sum_{k=s+1}^{m+2} \frac{\tau_n^k}{k!}B^{k-s-1}A.$$

For the proof, we can use mathematical induction, see [54].

So, for each further iteration step, we conserve the order $\mathcal{O}(\tau_n^{m+1})$ of Equation (2.67) or $\mathcal{O}(\tau_n^{m+2})$ of Equation (2.68).

We assume all local errors to be of order $\mathcal{O}(\tau_n^{m+2})$.

Based on this assumption, we obtain for the global error

$$\mathrm{err}_{\mathrm{global}}(t^{n+1}) = (n+1)\,\mathrm{err}_{\mathrm{local}}(\tau_n) = (n+1)\,\tau_n\,\frac{\mathrm{err}_{\mathrm{local}}(\tau_n)}{\tau_n} \tag{2.69}$$

$$= \mathcal{O}(\tau_n^{m+1}),$$

where we assume equidistant time steps, a time $t^{n+1} = (n+1)\,\tau_n$, and the same local error for all $n+1$ time steps, see also [153]. □

Remark 2.23. The exact solution of the split subproblem can also be extended to singular perturbed problems and unbounded operators. In these cases, the formal solution with respect to the asymptotic convergence of a power series, which is close to the exact solution, can be sought, see [9], [10].

Consistency Analysis of Iterative Splitting Method with Interpolated Split Solutions

The algorithm (2.40)–(2.41) requires the knowledge of the functions $u_{i-1}(t)$ and $u_i(t)$ over the whole interval $[t^n, t^{n+1}]$. However, when we solve the split subproblems, we usually apply some numerical methods that allow us to know the values of the above functions only at some points on the interval. Hence, typically we can define only some interpolations of the exact functions.

In the following, we consider and analyze the modified iterative process

$$\frac{\partial u_i(t)}{\partial t} = Au_i(t) + Bu_{i-1}^{\text{int}}(t), \text{ with } u_i(t^n) = u_{\text{sp}}^n, \qquad (2.70)$$

$$\frac{\partial u_{i+1}(t)}{\partial t} = Au_i^{\text{int}}(t) + Bu_{i+1}(t), \text{ with } u_{i+1}(t^n) = u_{\text{sp}}^n, \quad (2.71)$$

where $u_k^{\text{int}}(t)$ (for $k = i-1, i$) denotes an approximation of the function $u_k(t)$ on the interval $[t^n, t^{n+1}]$ with accuracy $\mathcal{O}(\tau_n^p)$. (For simplicity, we assume the same order of accuracy as the order p on each subinterval.)

Then, we lead to the following theorem:

Theorem 2.24. *Let $A, B \in \mathcal{L}(X)$ be given linear bounded operators and consider the abstract Cauchy problem (2.42). Then for any interpolation of order $p \geq 1$, the iteration (2.70)–(2.71) for $i = 1, 3, \ldots 2m + 1$ is consistent with the order of consistency α where $\alpha = \min\{2m - 1, p\}$.*

Proof. The iteration (2.70)–(2.71) for the error function $\mathcal{E}_i(t)$ recalls relation (2.46) with a modified right side, namely

$$\mathcal{F}_i(t) = \begin{bmatrix} Be_{i-1}(t) + Bh_{i-1}(t) \\ Ah_i(t) \end{bmatrix}, \qquad (2.72)$$

where $h_k(t) = u_k(t) - u_k^{\text{int}}(t) = \mathcal{O}(\tau_n^p)$ for $k = i-1, i$. Hence,

$$\|\mathcal{F}_i\|_\infty \leq \max\{\|B\| \, \|e_{i-1}\| + \|h_{i-1}\|; \|A\| \, \|h_i\|\}, \qquad (2.73)$$

which results in the estimation

$$\|\mathcal{F}_i\|_\infty \leq \|B\| \, \|e_{i-1}\| + C \, \tau_n^p. \qquad (2.74)$$

Consequently, based on these assumptions, estimation (2.58) turns into the following:

$$\|e_i\| \leq K(\|B\|\tau_n\|e_{i-1}\| + C \, \tau_n^{p+1}) + \mathcal{O}(\tau_n^2). \qquad (2.75)$$

Therefore, based on these assumptions, estimation (2.60) takes the modified form

$$\|e_{i+1}\| \leq K_1\tau_n^2\|e_{i-1}\| + KC\tau_n^{p+2} + KC\tau_n^{p+1} + \mathcal{O}(\tau_n^3), \qquad (2.76)$$

where $K, K_1 \in \mathbb{R}^+$ are constants and are not depending on the time step τ_n.

So we obtain the order of consistency $\alpha \min 2m - 1, p$, while p is the order of the interpolation and $2m - 1$ are the iterative steps. $\qquad \square$

Remark 2.25. Theorem 2.24 shows that the number of iterations should be chosen according to the order of the interpolation formula. For more iterations, we expect a more accurate solution.

Remark 2.26. As a result, we can use a piecewise constant approximation of the function $u_k(t)$, namely $u_k^{\text{int}}(t) = u_k(t^n) = \text{const}$, which is known from the split solution. In this instance, it is enough to perform only two iterations in the case of a sufficiently small discretization step size.

Remark 2.27. The above analysis was performed on the local error. The analysis of global error is the same as usual and leads to α-order convergence.

In the next section, we deal with the stability of time integration and time discretization methods.

2.6 Stability Analysis of Iterative Splitting Method for Bounded Operators

For the time integration methods, the stability is defined by the boundedness of the operators. Here, we have assumed that the exact underlying differential equations are to be solved.

The more realistic case in most applications is to use time integration or time discretization methods to solve the differential Equations (2.40)–(2.41).

Such methods are also time-consuming, if we assume that the solution is to be found with an accuracy at least as good as the underlying splitting scheme.

One of the benefits of dealing with implicit and explicit parts in the scheme (2.40)–(2.41) is to solve the explicit or fixed part directly and the implicit part with a given scheme.

So at least the more stiff operator is solved implicitly, while the nonstiff operator is solved explicitly. This means that we solve only the stiff part, while the nonstiff part can be solved directly or in less computational time.

For the time discretization methods, the stability has to be discussed in connection with the underlying discretization schemes and taking into account the different amounts of computational time needed by operators A and B.

We discuss the following schemes:

- Time integration schemes

- Time discretization schemes

Remark 2.28. In the first case, we assume we can directly integrate the ordinary differential equations and deal with exp-functions. In the other case we assume we wish to avoid exp-functions and apply time discretization methods.

For both cases, we discuss examples of their underlying methods, which are used later in the experiments, see Chapter 6.

2.6.1   Time Integration Methods

We concentrate on an approximation to the solution of the linear evolution equation

$$\partial_t\, u = Lu = (A + B)u,\ u(t^n) = u^n, \tag{2.77}$$

where $L, A, B \in \mathbb{R}^{m \times m}$ are bounded operators and m is a positive integer (rank of matrix).

2.6.1.1   Case 1: Alternating A and B

In this case, we assume we must solve alternating the initial value problem for A and B and can easily compute $\exp(At)$ and $\exp(Bt)$, while $\exp((A+B)t)$ is expensive.

As numerical method, we will imply a two-stage iterative splitting scheme:

$$u_i(t) \quad = \quad \exp(At)u^n + \int_{t^n}^t \exp(As)Bu_{i-1}(s)\, ds, \tag{2.78}$$

$$u_{i+1}(t) \quad = \quad \exp(Bt)u^n + \int_{t^n}^t \exp(Bs)Au_i(s)\, ds, \tag{2.79}$$

where $i = 1, 3, 5, \ldots$ and $u_0(t) = 0$ or $u_0(t) = u(t^n)$.

Based on the order of the iterative splitting scheme, we must take into account the same order of the time-integration schemes.

Evaluation with Trapezoidal Rule (Two Iteration Steps)

We must evaluate:

$$u_2(t) \quad = \quad \exp(Bt)u(t^n) \tag{2.80}$$

$$+ \quad \int_{t^n}^t \exp(B(t - s))Au_1(s)ds, \quad t \in (t^n, t^{n+1}],$$

where $u_1(t) = \exp(At)u(t^n)$ is the previous solution.

We apply the Trapezoidal rule and obtain:

$$u_2(t) = \exp(B\tau_n)u(t^n) + \tfrac{1}{2}\tau_n\left(Au_1(t) + \exp(B\tau_n)Au_1(t^n)\right), \tag{2.81}$$

where $\tau_n = t - t^n$.

Based on the bounded operators and the boundedness of the underlying exp-functions, the conditions for the stability of the method are given, see [117] and [124]. We obtain a second-order scheme given by two iterative steps and the Trapezoidal rule, which is a second-order scheme.

Evaluation with Simpson Rule (Three Iteration Steps)

We must evaluate:

$$u_3(t) \quad = \quad \exp(A\tau_n)u(t^n) \tag{2.82}$$
$$+ \quad \int_{t^n}^{t} \exp(A(t-s))Bu_2(s)ds, \quad t \in (t^n, t^{n+1}],$$

where $u_2(t)$ is given by the previous two-step method.

We apply the Simpson rule and obtain:

$$u_3(t) \quad = \exp(A\tau_n)u(t^n) \tag{2.83}$$
$$+ \frac{1}{6}\tau_n \left(Bu_2(t) + 4\exp(A\frac{\tau_n}{2})Bu_2((t+t^n)/2) + \exp(A\tau_n)Bu_2(t^n) \right),$$

where $\tau_n = t - t^n$.

Based on the bounded operators and the boundedness of the underlying exp-functions, the conditions for the stability of the method are given, see [117] and [124]. We obtain a third-order scheme given by three iteration steps and the Simpson rule, which is a third-order scheme.

Remark 2.29. The same result can also be derived by applying BDF3 (Backward Difference Formula of Third Order).

Evaluation with Bode Rule (Four Iteration Steps).

We must evaluate:

$$u_4(t) \quad = \quad \exp(B\tau_n)u(t^n) \tag{2.84}$$
$$+ \quad \int_{t^n}^{t} \exp(B(t-s))Au_3(s)ds, \quad t \in (t^n, t^{n+1}],$$

where $u_3(t)$ has to be evaluated with a third-order method.

We apply the Bode rule and obtain:

$$u_4(t) \quad = \quad \exp(B\tau_n)u(t^n) \tag{2.85}$$
$$+ \quad \frac{1}{90}\tau_n \left(7Au_3(t) + 32\exp(B\frac{\tau_n}{4})Au_3(\frac{3t+t^n}{4}) \right.$$
$$+ \quad 12\exp(B\frac{\tau_n}{2})Au_3(\frac{t+t^n}{2}) + 32\exp(B\frac{3\tau_n}{4})Au_3(\frac{t+3t^n}{4})$$
$$+ \quad \left. 7\exp(B\tau_n)Au_3(t^n) \right),$$

where $\tau_n = t - t^n$ and $u_3(t)$ is evaluated with the Simpson rule or another third-order method.

Based on the bounded operators and the boundedness of the underlying exp-functions, the conditions for the stability of the method are given, see [117] and [124]. We obtain a fourth order scheme given by four iteration steps and the Bode rule, which is a third-order scheme.

Remark 2.30. The same result can also be derived by applying the fourth order Gauss Runge-Kutta method. We can generalize the schemes by using more iteration steps and higher-order time integration schemes.

2.6.1.2   Case 2: A is Stiff and B is Nonstiff

In this case, we assume we must solve the initial value problem only over A and B is explicit. We assume we must compute only $\exp(At)$, while B is only a perturbation.

As a numerical method, we imply a one-stage iterative splitting scheme:

$$u_i(t) = \exp(At)u^n + \int_{t^n}^{t} \exp(As)Bu_{i-1} \, ds, \qquad (2.86)$$

where $i = 1, 2, 3, \ldots$ and $u_0(t) = 0$ or $u_0(t) = u(t^n)$.

Based on the order of the one-stage iterative splitting scheme, we have to take into account the same order in the time integration schemes.

Evaluation with Trapezoidal Rule (Two Iteration Steps)

We must evaluate:

$$u_2(t) = \exp(At)u(t^n) + \int_{t^n}^{t} \exp(A(t - s))Bu_1(s)ds, \quad t \in (t^n, t^{n+1}], \quad (2.87)$$

where $u_1(t) = \exp(At)u(t^n)$ is the previous solution.

We apply the Trapezoidal rule and obtain:

$$u_2(t) = \exp(A\tau_n)u(t^n) + \tfrac{1}{2}\tau_n \left(Bu_1(t) + \exp(A\tau_n)Bu_1(t^n)\right), \qquad (2.88)$$

where $\tau_n = t - t^n$.

Based on the bounded operators, the method is stable, see also the arguments in Case 1. We obtain a second-order scheme given by two iteration steps and the Trapezoidal rule, which is a second-order scheme.

Evaluation with Simpson Rule (Three Iteration Steps)

We must evaluate:

$$
\begin{aligned}
u_3(t) \;=\;& \exp(A\tau_n)u(t^n) \\
&+ \int_{t^n}^{t} \exp(A(t - s))Bu_2(s)ds, \quad t \in (t^n, t^{n+1}],
\end{aligned}
\qquad (2.89)
$$

where $u_2(t)$ is given by the previous two-step method.

We apply the Simpson rule and obtain:

$$
\begin{aligned}
u_3(t) \;=\; & \exp(A\tau_n)u(t^n) \hspace{4cm} (2.90)\\
+\; & \frac{1}{6}\tau_n\left(Bu_2(t) + 4\exp\!\left(A\frac{\tau_n}{2}\right)Bu_2((t+t^n)/2)\right.\\
+\; & \left. \exp(A\tau_n)Bu_2(t^n)\right),
\end{aligned}
$$

where $\tau_n = t - t^n$.

Based on the bounded operators, the method is stable, see also the arguments in Case 1. We obtain a third-order scheme given by three iteration steps and the Simpson rule, which is a third-order scheme.

Remark 2.31. The same result can also be derived by applying BDF3 (Backward Difference Formula of Third Order).

Evaluation with Bode Rule (Four Iteration Steps)

We must evaluate:

$$
\begin{aligned}
u_4(t) \;=\; & \exp(A\tau_n)u(t^n) \hspace{4cm} (2.91)\\
+\; & \int_{t^n}^{t} \exp(A(t-s))Bu_3(s)ds, \quad t \in (t^n, t^{n+1}],
\end{aligned}
$$

where $u_3(t)$ has to be evaluated with a third-order method.

We apply the Bode rule and obtain:

$$
\begin{aligned}
u_4(t) \;=\; & \exp(A\tau_n)u(t^n) \hspace{4cm} (2.92)\\
+\; & \frac{1}{90}\tau_n\left(7Bu_3(t) + 32\exp\!\left(A\frac{\tau_n}{4}\right)Bu_3\!\left(\frac{3t+t^n}{4}\right)\right.\\
+\; & 12\exp\!\left(A\frac{\tau_n}{2}\right)Bu_3\!\left(\frac{t+t^n}{2}\right) + 32\exp\!\left(A\frac{3\tau_n}{4}\right)Bu_3\!\left(\frac{t+3t^n}{4}\right)\\
+\; & \left. 7\exp(A\tau_n)Bu_3(t^n)\right),
\end{aligned}
$$

where $\tau_n = t - t^n$ and $u_3(t)$ is evaluated with the Simpson rule or another third-order method.

Based on the bounded operators, the method is stable, see also the arguments in Case 1. We obtain a fourth-order scheme given by four iteration steps and the Bode rule, which is a third-order scheme.

Remark 2.32. The same result can also be derived by applying the fourth-order Gauss Runge-Kutta method. We can generalize the schemes by using more iteration steps and higher-order time integration schemes.

Remark 2.33. We can generalize our results to higher-order time integration methods for bounded operators. Based on the assumptions, that all operators and their underlying integrals in the schemes (2.78)–(2.79) and (2.86) are bounded in a Banach norm, the underlying time integration methods are at least stable; see the ideas of a proof in [86] and [106].

2.6.2   Time Discretization Methods

We deal with time discretization methods used for discretizing the iterative splitting scheme. The methods and proof ideas are discussed in the preprint of Geiser/Tanoglu [98].

For the application to numerical experiments, we assume that the iterative splitting method (2.40)–(2.41) will be applied to linear evolution systems, e.g., diffusion-reaction systems.

We assume we will apply semidiscretization to the spatially dependent operators, e.g., diffusion and convection operator is discretized with finite difference methods, see [118], and obtain the matrices A and B. Further, we assume that we can estimate the resulting matrices in a Banach norm.

Based on this ordinary differential equation, we apply time discretization. We make the following assumptions:

Assumption 2.34. *The time discretization methods are applied to the iterative splitting method (2.40)–(2.41). We consider some suitable vector norm $||\cdot||$ on $\mathbb{R}^M$ and the corresponding induced matrix norm $||\cdot||$ on $\mathbb{R}^{M \times M}$. The matrix exponential of $Z \in \mathbb{R}^{M \times M}$ is denoted by $\exp(Z)$.*

- *A and B are matrices, they can be the result of a finite difference discretization or of the ordinary differential equations given.*

- *Both A and B have only real negative eigenvalues.*

- *For operators with real negative eigenvalues, we obtain:*

$$|| \exp(\tau_n A)|| \leq 1, || \exp(\tau_n B)|| \leq 1 \, , \forall \tau_n > 0. \tag{2.93}$$

- *If A and B are discretized spatial derivative operators, then Au_0 and Bu_0 will be bounded if the initial solution u_0 is smooth and satisfies the appropriate boundary condition for $t = 0$.*

- *The implicit Euler and Crank-Nicolson method is unconditionally stable:*

$$||R_{CN}(Z)|| = ||(I - 1/2Z)^{-1}(I + 1/2Z)|| \leq 1, \tag{2.94}$$

$$||R_{impl.Eu}(Z)|| = ||(I - Z)^{-1}|| \leq 1, \tag{2.95}$$

where $Z = \tau_n A$ or $Z = \tau_n B$ and R_{CN}, $R_{impl.Eu}$ are the transfer functions for the A-stable scheme.

For time discretization, we propose the Runge-Kutta and BDF methods.

Even for real eigenvalues, the instabilities that are often observed can be overcome with some assumptions about the operators.

In this way, we can stabilize the iterative splitting scheme with Runge-Kutta and BDF methods.

Remark 2.35. In the following, we discuss the two-stage iterative splitting method, which is assumed to give some benefit when decoupling the operators A and B.

2.6.2.1 Runge-Kutta Methods

We discuss the stabilization of the iterative splitting method with Runge-Kutta methods of first and second orders. Here we deal with explicit and implicit components of the splitting scheme. The implicit discretization method is used for the operators that have been solved with an implicit method, while the explicit discretization method is used for all the operators that can be done with the previous iterative solution. Here the balance of these discretization schemes stabilize the iterative splitting scheme.

Theorem 2.36. *Let A and B be linear and bounded operators in a Banach space $\mathbf{X} = \mathbb{R}^M$ according to Assumption 2.34.*

We apply implicit and explicit Euler and Crank-Nicolson methods to obtain stable time discretizations of the iterative splitting scheme (2.40)–(2.41).

Then the amplification matrix of iterative splitting fulfills unconditional stability:

$$||R(Z_1, Z_2)|| \leq 1 \, , \forall \tau_n > 0, \tag{2.96}$$

where $Z_1 = \tau_n A$ and $Z_2 = \tau_n B$ and R is the transfer function of the discretization scheme.

Proof. In the discrete case, we can balance the implicit and explicit schemes, such that we obtain stabile methods. We assume two stages for the iterative method and discretize with a θ-method:

$$\begin{aligned}
\overline{u}_{i+1}^{n+1} &= u_i^n + \tau_n(1 - \theta_1)(A(u_{i+1}^n) + B(u_i^n)) \\
&\quad + \tau_n \theta_1(A(\overline{u}_{i+1}^{n+1}) + B(u_i^{n+1})) \,, \tag{2.97} \\
u_{i+1}^{n+1} &= u_{i+1}^n + \tau_n(1 - \theta_2)(A(u_{i+1}^n) + B(u_{i+1}^n)) \\
&\quad + \tau_n \theta_2(A(\overline{u}_{i+1}^{n+1}) + B(u_{i+1}^{n+1})) \,, \tag{2.98}
\end{aligned}$$

where $u_i^n = u_{i+1}^n = u^n$ and the initialization is with $u_0^{n+1} = u^n$

For the linear system, we denote $Z_1 = \tau_n A$ and $Z_2 = \tau_n B$ and we set $\theta_1 = \theta_2$.

We get the following stability equation, see [134], and for $\theta = 1/2$, we compute the first iteration with $i = 1$ and get the equation

$$
\begin{aligned}
u_1^{n+1} &= (I + (I - 1/2Z_2)^{-1}(I - 1/2Z_1)^{-1}(Z_1 + Z_2)u^n , & (2.99) \\
&= (I - 1/2Z_2)^{-1}((I - 1/2Z_2) + (I - 1/2Z_1)^{-1}(Z_1 + Z_2))u^n \\
&= (I - 1/2Z_2)^{-1}(I - 1/2Z_1)^{-1}(I + 1/2Z_1)(I + 1/2Z_2)u^n
\end{aligned}
$$

Setting $\tilde{u}^n = (I - 1/2Z_2)u^n$, we obtain:

$$
\begin{aligned}
\tilde{u}^{n+1} &= (I - 1/2Z_1)^{-1}(I + 1/2Z_1) & (2.100) \\
&\quad \cdot (I + 1/2Z_2)(I - 1/2Z_2)^{-1}\tilde{u}^n \\
&= R_{CN}(1/2Z_1)R_{CN}(1/2Z_2)\tilde{u}^n
\end{aligned}
$$

where, according to the inner product norm and Assumption 2.34, it follows that the stability functions of R_{CN} are all bounded by 1.

Hence, we have:

$$
||\tilde{u}^{n+1}|| \leq ||\tilde{u}^n||, \tag{2.101}
$$

and

$$
||u^n|| \leq ||(I - \frac{1}{2}Z_2)u_0||, \ \forall n \geq 0. \tag{2.102}
$$

This gives stability results with respect to $||Z_2 u_0|| \leq C||u_0||$.

This is fulfilled by our Assumption 2.34, that Bu_0 is bounded.

$\square$

Remark 2.37. Here we have the benefit in decoupling into two equations and could deal with different timesteps. So we can reduce the computational time and increase the timestep in the less stiff part.

2.6.2.2   BDF Methods

In this section, we derive the stability of backward difference formulae (BDFk), where k denotes the order of the method, in each subequation of the iterative splitting scheme (2.40)–(2.41).

Introduction to the BDFk Methods

In the literature, the BDF is given by the equation,

$$
\sum_{r=1}^{k+1} \alpha_r u^{n-r+2} = \tau_n \beta f(u^{n+1}). \tag{2.103}
$$

where the coefficients α_r and β are obtained from a Taylor expansion and the difference operator. We apply the BDF method of order f for the operators A and B and obtain:

$$\sum_{r=1}^{k+1} \alpha_r u^{n-r+2} = \tau_n \beta A u^{n+1} + \tau_n \beta B u^{n+1}, \qquad (2.104)$$

and the following conditions must be satisfied:

1. $k < 7$, otherwise the method is not zero stable,

2. $\sum_{r=1}^{k+1} \alpha_r = 0$, otherwise the method is not consistent.

The construction of the BDFk methods are given in [123].

Stability Proofs

For the following theorems, we consider some suitable vector norm $|| \cdot ||$ on $\mathbb{R}^m$, together with a suitable indexed operator norm.

The matrix exponential of $Z \in \mathbb{R} \times \mathbb{R}$ will be denoted by $\exp(Z)$ and we assume that we have dissipative operators A, B with

$$|| \exp(\tau_n A) || \le 1, || \exp(\tau_n B) || < 1, \text{ for all } \tau_n > 0. \qquad (2.105)$$

This means that the operators are themselves stable.

We have the following assumptions:

Assumption 2.38. *The time discretization methods are applied to the iterative operator-splitting method (2.40)–(2.41). We consider some suitable vector norm $|| \cdot ||$ on $\mathbb{R}^M$. The matrix exponential of $Z \in \mathbb{R}^{M \times M}$ is denoted by $\exp(Z)$.*

- *A and B are matrices, for example given by spatial dicretized operators and they define the ordinary differential equations.*

- *Both A and B have real nonpositive eigenvalues*

- *We assume:*

$$|| \exp(\tau_n A) || \le 1, || \exp(\tau_n B) || \le 1, \forall \tau_n > 0, \qquad (2.106)$$

- *If A and B are discretized spatial derivative operators, then $A u_0$ and $B u_0$ will be bounded if the initial solution u_0 is smooth and satisfies the appropriate boundary condition for $t = 0$.*

- *The implicit Euler and Crank-Nicolson method is unconditionally stable:*

$$||R_{CN}(Z)|| = ||(I - 1/2Z)^{-1}(I + 1/2Z)|| \le 1, \qquad (2.107)$$

$$||R_{impl.Eu}(Z)|| = ||(I - Z)^{-1}|| \le 1, \qquad (2.108)$$

where $Z = \tau_n A$ or $Z = \tau_n B$.

- *We assume that the initial solution is sufficiently smooth:*
  for $i = 1$: $||Bu^n|| \leq C_1||u^n||$,
  for $i = 2$: $||ABu^n|| \leq C_2||u^n||$.

Then we can generalize the stability function according to the following theorem:

Theorem 2.39. *We apply BDF methods to the iterative splitting method given in Equations (2.40)–(2.41), then for the linear system with $Z_1 = \tau_n A$ and $Z_2 = \tau_n B$ we obtain stable functions $R_i(Z_1, Z_2)$ under Assumption 2.38, that are given as*

$$||R_i(Z_1, Z_2)|| \leq 1, \quad \text{for all } Z_1, Z_2 \in \mathbf{X} \times \mathbf{X}, \qquad (2.109)$$

where $\mathbf{X}$ is a Banach-space, $|| \cdot ||$ is the matrix norm and $i = 1, 2$ and R_i is the i-th transfer function of the splitting scheme.

Such that the iterative splitting methods are stable with the underlying BDF methods as time discretization methods.

Proof. Discretization of Equations (2.40)–(2.41) by BDFk methods gives:

$$(\alpha_1 I - \beta\tau_n A)u_i^{n+1} = -\sum_{r=2}^{k+1} \alpha_k u_i^{n-r+2} + \beta\tau_n B u_{i-1}^{n+1} \qquad (2.110)$$

$$(\alpha_1 I - \beta\tau_n B)u_{i+1}^{n+1} = -\sum_{r=2}^{k+1} \alpha_k u_i^{n-r+2} + \beta\tau_n A u_i^{n+1}, \qquad (2.111)$$

where $i = 1, 3, 5, \ldots$.

For the linear system, we denote $Z_1 = \tau_n A$ and $Z_2 = \tau_n B$ and we set $u_i^n = u_{i+1}^n = u_i^{n-1} = u_{i+1}^{n-1} = \ldots = u_i^{n-k+1} = u_{i+1}^{n-k+1} = u^n$ and initialize with $u_0^{n+1} = u^n$.

We get the following stability equation by following the same approach given in reference [133].

We compute the first iteration with $i = 1$ and get the following equation

$$(\alpha_1 I - \beta Z_1)u_1^{n+1} = -\sum_{r=2}^{k+1} \alpha_i u^n + \beta Z_2 c^n, \qquad (2.112)$$

$$= \alpha_1 u^n + \beta Z_2 c^n,$$

$$= (\alpha_1 I + \beta Z_2)u^n,$$

$$u_1^{n+1} = (I - a_k Z_1)^{-1}(I + a_k Z_2)u^n, \qquad (2.113)$$

$$u_1^{n+1} = R_1(Z_1, Z_2)u^n,$$

where $a_k = \frac{\beta}{\alpha_1}$.

Setting

$$\tilde{u}^n = (I - a_k Z_2)u^n, \qquad (2.114)$$

we obtain:

$$\tilde{u}^{n+1} = (I - a_k Z_2)(I - a_k Z_1)^{-1}(I + a_k Z_2)(I - a_k Z_2)^{-1}\tilde{u}^n \quad (2.115)$$
$$= (I - a_k Z_2)R_{im.Eu.}(a_k Z_1)R_\theta(a_k Z_2)\tilde{u}^n$$

where, with the inner product norm and Assumption 2.38, it follows that the stability functions of $R_{im.Eu.}$ and R_θ are all bounded by 1.

We obtain:

$$||\tilde{u}^n|| \leq ||(I - a_k Z_2)^n \tilde{u}^0||, \forall n \geq 0. \quad (2.116)$$

Further, we reset the setting of 2.114

$$||u^n|| \leq ||(I - a_k Z_2)^{n+1} u^0||, \forall n \geq 0, \quad (2.117)$$

and assume $||Bu^n|| \leq C||u^n||, \forall n \geq 0$, such that we have stable results.

This also gives stability to the spatial derivative operators, if the initial solution is smooth and satisfies the appropriate boundary conditions.

For the next iteration $i = 2$, we get the same stability function as before

$$(\alpha_1 I - \beta Z_2)u_2^{n+1} = -\sum_{i=2}^{k+1} \alpha_i u^n + \beta Z_1 u_1^{n+1}, \quad (2.118)$$
$$= \alpha_1 u^n + \beta Z_1 R_1(Z_1, Z_2)u^n,$$
$$= (\alpha_1 I + \beta Z_1 R_1(Z_1, Z_2))u^n,$$
$$u_2^{n+1} = (I - a_k Z_2)^{-1}(I + a_k R_1(Z_1, Z_2)Z_1)c^n, \quad (2.119)$$
$$u_2^{n+1} = (I - a_k Z_2)^{-1}(I - a_k Z_1)^{-1}(I + a_k^2 Z_1 Z_2)c^n,$$
$$= R_2(Z_1, Z_2)u^n$$

so we obtain,

$$u_2^{n+1} = R_{im.Eu.}(a_k Z_2)R_{im.Eu.}(a_k Z_1)(I + a_k^2 Z_1 Z_2)u^n, \quad (2.120)$$

We derive the stability as

$$||u_2^n|| \leq ||(I + a_k^2 Z_1 Z_2)^n u^0||, \forall n \geq 0, \quad (2.121)$$

and assume $||ABu^n|| \leq C||u^n||, \forall n \geq 0$ and have stable results.

□

Remark 2.40. A generalization for $i > 2$ is also possible. The underlying idea of the proof is to derive an induction with boundable stability functions of the previous iterative step, see [98]. The main assumption is to have sufficiently smooth initial solutions to bound the operators. For example, for $i = 3$, we have to bound $BABu_0$ with sufficient smoothness of the initial condition and satisfy the appropriate boundary condition.

Remark 2.41. The flexibility in choosing the iterative steps over operator A or B obtains a scheme that takes into account the best choice of the smallest norms of A or B. This means we iterate over A if the norm of B is smaller or vice versa. Here the benefit of balancing implicit and explicit parts are given, while the operator with the larger norm is computed implicit and the operator with the smaller norm is computed explicit; see also numerical Benchmarks in Chapter 6.

Chapter 3

Decomposition Methods for Partial Differential Equations

In this chapter we concentrate on decoupling differential equations with unbounded operators.

We consider the abstract Cauchy problem in a Banach space $\mathbf{X}$

$$\partial_t c(x, t) = Ac(x, t) + Bc(x, t), \quad 0 < t \leq T, \tag{3.1}$$
$$c(0) = c_0,$$

where $A : dom(A) \to \mathbf{X}, B : dom(B) \to \mathbf{X}$ and $dom(A) \subset \mathbf{X}, dom(B) \subset \mathbf{X}$ are given linear operators that are generators of the analytical semigroups and $c_0 \in \mathbf{X}$ is a given element. We assume an operator norm to the Banach space given as $|| \cdot ||$.

We extend our theory to unbounded operators, because also semidiscretized spatial depending operators are at least unbounded, see [180].

Instead of the bounded operators, we have to extend to analytical semigroups, see [181], to estimate the integral formulations in a related operator norm $|| \cdot ||$.

We concentrate on initial value problems, meaning Cauchy problems, where the problem can result of an ODE system or we assume to have discretized partial differential equations in space with finite difference methods and the boundary values are included. We assume that these problems lead to an initial value problem with linear unbounded operators, see [180].

We have the following cases:

- Both operators have the same domains $dom(A) = dom(B)$, meaning we concentrate on operators that only differ to some constants.

- We assume to different operators, $dom(A) \subset dom(B)$ and we denote:

$$B = A^{1-\alpha}, \tag{3.2}$$

  where $\alpha \in (0, 1)$.

- Extension can be done for:

$$B = f(A), \tag{3.3}$$

  where $f : \mathbb{R} \to \mathbb{R}$ is a polynomial function.

- For the noncommutative case, we can also extend to:

$$B \leq f(A), \tag{3.4}$$

where $f : \mathbb{R} \to \mathbb{R}$ is a polynomial function.

We concentrate on the cases for the commuting operators and $B = A^{1-\alpha}$.

The extension can be treated in a similar way; see ideas for application to parabolic initial boundary value problems [181].

The idea of the proof techniques are given in the following Figure 3.1.

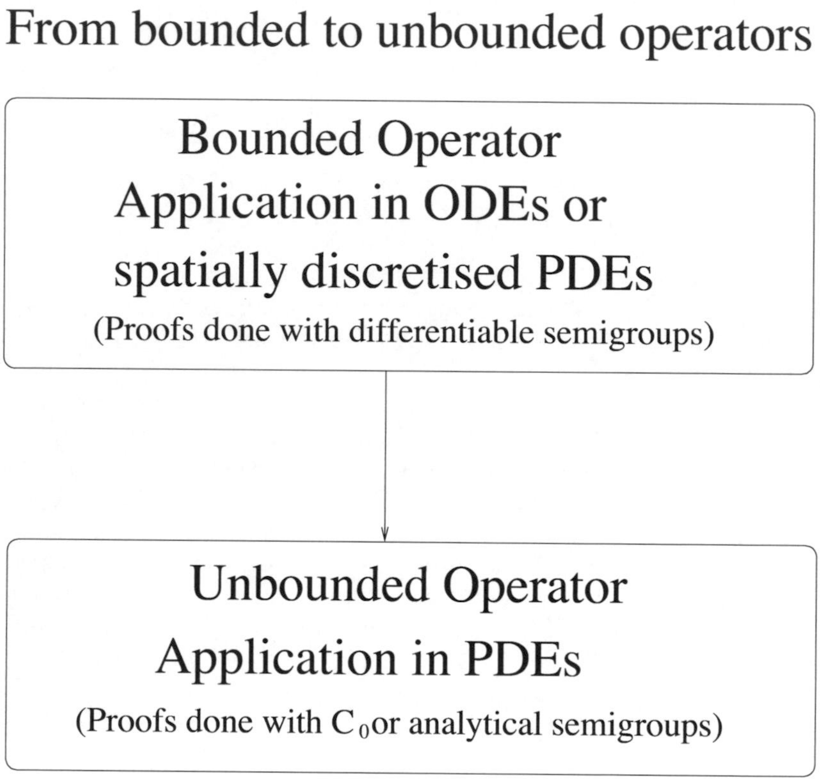

From bounded to unbounded operators

FIGURE 3.1: Proof techniques for unbounded operators.

3.1   Iterative Schemes for Unbounded Operators

We discuss the following based on the different assumptions to the operators two iterative schemes.

While one operator is at least less bounded that the other operator, we reduce the order or leave the same order if we are iterative over the less bounded operator.

3.1.1 Iterative Splitting Schemes

In the following we deal with the following scheme:
1. One-stage iterative splitting method

$$u_i(t) = \exp(At)u_0 + \int_0^t \exp(A(t-s))Bu_{i-1}(s)\, ds, \qquad (3.5)$$

$$u_i(0) = u(0), \qquad (3.6)$$

where $i = 1, 2, 3, \ldots$ and $u_0(t) = 0$.
2. Two-stage iterative splitting method

$$u_i(t) = \exp(At)u_0 + \int_0^t \exp(A(t-s))Bu_{i-1}(s)\, ds, \qquad (3.7)$$

$$u_i(0) = u(0), \qquad (3.8)$$

$$u_{i+1}(t) = \exp(Bt)u_0 + \int_0^t \exp(B(t-s))Au_i(s)\, ds, \qquad (3.9)$$

$$u_{i+1}(0) = u(0), \qquad (3.10)$$

where $i = 1, 3, 5, \ldots$ and $u_0(t) = 0$.
We deal with the following assumptions:

Assumption 3.1. *The linear operators $A + B, A, B$ generate C_0 semigroups on $\mathbf{X}$, and the operators A, B satisfy in addition the bounds:*

$$\|\exp(A\tau)\| \le \exp(\omega|t|) \ and \ \|\exp(B\tau)\| \le \exp(\omega|t|) \qquad (3.11)$$

for some $\omega \ge 0$ and all $t \in \mathbb{R}$.

The following theorem is given the convergence of an iterative operator splitting scheme for one-sided iterations and we assume that exists $B = A^{1-\alpha}$:

Assumption 3.2. *For the consistency proofs we have to assume the following:*
The linear operators $A + B, A, B$ generate analytical semigroups on $\mathbf{X}$, and the operators A, B satisfy in addition the bounds:

$$\|B^\alpha \exp(B\tau_n)\| \le \kappa_1 \tau_n^{-\alpha}. \qquad (3.12)$$

$$\|B \exp((A+B)\tau_n)\| \le \kappa_2 \tau_n^{-1+\alpha}, \qquad (3.13)$$

$$\|\exp(A\tau_n)B\| \le \kappa_3 \tau_n^{-1+\alpha}, \qquad (3.14)$$

$$\|A^\beta \exp(A\tau_n)\| \le \kappa_4 \tau_n^{-\beta}. \qquad (3.15)$$

$$\|A^\gamma \exp((A+B)\tau_n)\| \le \kappa_5 \tau_n^{-\gamma}, \qquad (3.16)$$

where $\alpha, \beta, \gamma \in (0, 1)$, $\tau_n = (t^{n+1} - t^n)$ and κ_i for $i = 1, \ldots, 5$ are constants, see [180].

Remark 3.3. For the one-stage iterative scheme it is sufficient to have the assumptions: (3.13)–(3.16), where for the two-stage iterative scheme we need all assumptions as for the one-stage schemes and additionally assumption (3.12).

That means that we assume that operator A and B generates an analytical semigroup with $\exp(At)$ and $\exp(Bt)$.

3.1.2   One-Stage Iterative Splitting Method for A-bounded Operators

In the following we present the convergence of the one-stage iterative splitting method.

Theorem 3.4. *For the numerical solution of (2.40), consider an iterative operator splitting scheme on operator A with i-th iterative steps.*

If the Assumptions 3.1 and 3.2 are valid, then

$$\|S_i^n - \exp((A+B)n\tau)\| \leq C\tau^{i-1}, \ n\tau \leq T, \tag{3.17}$$

where the constant C can be chosen uniformly on bounded-time intervals and in particular, independent of n and τ.

Proof. By applying the telescopic identity we obtain

$$(S_i^n - \exp((A+B)n\tau)u_0 \tag{3.18}$$
$$= \sum_{v=0}^{n-1} S_i^{n-v-1}(S - \exp((A+B)\tau)) \exp(v\tau(A+B))u_0,$$

if we assume the stability bound:

$$\|S_i\| \leq \exp(c\omega\tau), \tag{3.19}$$

with a constant c only depends on the estimation of the method.

Furthermore, if we assume the consistency bound:

$$\|S_i^n - \exp((A+B)n\tau)\|$$
$$\leq \exp(c\omega T) \sum_{v=0}^{n-1} \|(S - \exp(\tau(A+B))) \exp(v\tau(A+B))\| \tag{3.20}$$
$$\leq C\tau^{i-1}, \ n\tau \leq T. \tag{3.21}$$

The desired consistency and stability bound is given in the next subsubsections.

$\square$

3.1.2.1 Consistency Analysis

We present the results of the consistency of our iterative method. We assume for the system of operator the generator of an analytical semigroup based on their underlying norms for the Banach space $\mathbf{X}$ and induced operator norm denoted by $|| \cdot ||$.

In the following we discuss the consistency of the to-stage iterative method, taken into account to iterate over both operators.

Theorem 3.5. *Let us consider the abstract Cauchy problem in a Banach space* $\mathbf{X}$ *given in Equation (3.46). With the operators* $A, B : \mathbf{X} \to \mathbf{X}$ *are linear operators that are generators of the analytical semigroups. We assume* $dom(B) \subset dom(A)$*, so we are restricted to balance the operators. We assume*

$$B = A^{1-\alpha} \tag{3.22}$$

is the infinitesimal generator of an analytical semigroup for all $\alpha \in (0,1)$*, see [180].*

The consistency error is given as $\mathcal{O}(\tau_n^{i\alpha})$*, where* $\tau_n = t^{n+1} - t^n$ *and we have equidistant time steps, with* $n = 1, \ldots, N$*. Further* i *is the iterative step with operator* A*.*

Then the iteration process (3.5) for $i = 1, 2, 3, \ldots$ *is consistent with the order of the consistency* $\mathcal{O}(\tau_n^{\alpha i})$*, where* $0 \leq \alpha < 1$*.*

Proof. Let us consider the iteration (3.5) on the subinterval $[t^n, t^{n+1}]$.

For the first iteration we have:

$$\partial_t c_1(t) = A c_1(t), \quad t \in (t^n, t^{n+1}], \tag{3.23}$$

and for the second iteration we have:

$$\partial_t c_2(t) = A c_2(t) + B c_1(t), \quad t \in (t^n, t^{n+1}], \tag{3.24}$$

In general we have:

for $i = 1, 2, 3, \ldots$

$$\partial_t c_i(t) = A c_i(t) + B c_{i-1}(t), \quad t \in (t^n, t^{n+1}], \tag{3.25}$$

where for $c_0(t) \equiv 0$.

The solutions for the first two iterative steps are given by the variation of constants:

$$c_1(t) = \exp(A(t - t^n))c(t^n), \quad t \in (t^n, t^{n+1}], \tag{3.26}$$

$$
\begin{aligned}
c_2(t) \quad &= \exp(A(t - t^n))c(t^n) \\
&+ \int_{t^n}^{t^{n+1}} \exp(A(t^{n+1} - s))B c_1(s)ds, \quad t \in (t^n, t^{n+1}].
\end{aligned}
\tag{3.27}
$$

For the recursive iterations we have the solutions for $i = 1, 2, 3, \ldots$:

$$c_i(t) = \exp(A(t - t^n))c(t^n) + \int_{t^n}^t \exp((t - s)A)Bc_{i-1}(s) \, ds, \quad t \in (t^n, t^{n+1}], \tag{3.28}$$

The consistency is given in the following steps:

For e_1 we have:

$$c_1(t^{n+1}) = \exp(A\tau_n)c(t^n), \tag{3.29}$$

$$c(t^{n+1}) = \exp((A + B)\tau_n)c(t^n) = \exp(A\tau_n)c(t^n) \tag{3.30}$$
$$+ \int_{t^n}^{t^{n+1}} \exp(A(t^{n+1} - s))B \exp((s - t^n)(A + B))c(t^n) \, ds.$$

We obtain:

$$||e_1|| = ||c - c_1|| \leq || \exp((A + B)\tau_n)c(t^n) - \exp(A\tau_n)c(t^n)|| \tag{3.31}$$
$$\leq || \int_{t^n}^{t^{n+1}} \exp(A(t^{n+1} - s))B \exp((s - t^n)(A + B))c(t^n) \, ds||$$
$$\leq || \int_{t^n}^{t^{n+1}} \exp(A(t^{n+1} - s))A^{1-\alpha} \exp((s - t^n)(A + B))c(t^n) \, ds||$$
$$\leq \int_{t^n}^{t^{n+1}} Cs^{\alpha-1} ds||c(t^n)||$$
$$\leq C\tau^\alpha \, ||c(t^n)||$$

where $\alpha \in (0, 1)$, $\tau = (t^{n+1} - t^n)$ and C is a constant depends only on κ_5 and ω.

For e_2 we have:

$$c_2(t^{n+1}) = \exp(A\tau_n)c(t^n) \tag{3.32}$$
$$+ \int_{t^n}^{t^{n+1}} \exp(A(t^{n+1} - s))B \exp((s - t^n)A)c(t^n) \, ds,$$

$$c(t^{n+1}) = \exp(A\tau_n)c(t^n) \tag{3.33}$$
$$+ \int_{t^n}^{t^{n+1}} \exp(A(t^{n+1} - s))B \exp((s - t^n)A)c(t^n) \, ds$$
$$+ \int_{t^n}^{t^{n+1}} \exp(A(t^{n+1} - s))B$$
$$\int_{t^n}^s \exp(A(s - \rho))B \exp((\rho - t^n)(A + B))c(t^n) \, d\rho \, ds.$$

We obtain:

$$||e_2|| \leq ||\exp((A+B)\tau_n)c(t^n) - c_2|| \tag{3.34}$$

$$= ||\int_{t^n}^{t^{n+1}} \exp(A(t^{n+1} - s))B$$

$$\int_{t^n}^{s} \exp(A(s-\rho))B\exp((\rho - t^n)(A+B))c(t^n) \, d\rho \, ds||$$

$$= \int_{t^n}^{t^{n+1}} ||\exp(A(t^{n+1} - s))||$$

$$\int_{t^n}^{s} ||\exp(A(s-\rho))A^{2-2\alpha}\exp((\rho - t^n)(A+B))c(t^n) \, d\rho|| ds$$

$$= \int_{t^n}^{t^{n+1}} C \int_{t^n}^{s} (s-\rho)^{2\alpha-2} d\rho ds||c(t^n)||$$

$$\leq C\tau^{2\alpha} ||c(t^n)||$$

where $\alpha \in (0,1)$, $\tau = t^{n+1} - t^n$ and C is a constant only depending on κ_5 and ω.

For the general iterative steps, the recursive proof is given in the following. We shift $t^n \to 0$ and $t^{n+1} \to \tau_n$ for simpler calculations, see [136]. The initial conditions are given with $c(0) = c(t^n)$.

For the iterative steps $i = 1, 2, 3, \ldots$, we obtain for c_i and c:

$$c_i(\tau_n) = \exp(A\tau_n)c(0) \tag{3.35}$$

$$+ \int_0^{\tau_n} \exp(As)B\exp((\tau_n - s)B)c(0) \, ds$$

$$+ \int_0^{\tau_n} \exp(As_1)B \int_0^{\tau_n - s_1} \exp(s_2 B)A\exp((\tau_n - s_1 - s_2)A)c(0) \, ds_2 \, ds_1$$

$$+ \ldots +$$

$$+ \int_0^{\tau_n} \exp(As_1)B \int_0^{\tau_n - s_1} \exp(s_2 A)B \int_0^{\tau_n - s_1 - s_2} \exp(s_3 A)B \ldots$$

$$\int_0^{\tau_n - \sum_{j=1}^{i-1} s_j} \exp(As_i)B\exp((\tau_n - \sum_{j=1}^{i-1} s_j)A)c(0) \, ds_i \ldots ds_1,$$

$$c(\tau_n) = \exp(A\tau_n)c(0) \qquad (3.36)$$

$$+ \int_0^{\tau_n} \exp(As)B\exp((\tau_n - s)B)c(0)\,ds$$

$$+ \int_0^{\tau_n} \exp(As_1)B \int_0^{\tau_n - s_1} \exp(s_2 B)A\exp((\tau_n - s_1 - s_2)A)c(0)\,ds_2\,ds_1$$

$$+ \ldots +$$

$$+ \int_0^{\tau_n} \exp(As_1)B \int_0^{\tau_n - s_1} \exp(s_2 A)B \int_0^{\tau_n - s_1 - s_2} \exp(s_3 A)B \ldots$$

$$\int_0^{\tau_n - \sum_{j=1}^{i-1} s_j} \exp(As_i)B\exp\left((\tau_n - \sum_{j=1}^{i-1} s_j)A\right)c(0)\,ds_i \ldots ds_1$$

$$+ \int_0^{\tau_n} \exp(As_1)B \int_0^{\tau_n - s_1} \exp(s_2 A)B \int_0^{\tau_n - s_1 - s_2} \exp(s_3 A)B \ldots$$

$$\int_0^{\tau_n - \sum_{j=1}^{i} s_j} \exp(As_{i+1})B\exp\left((\tau_n - \sum_{j=1}^{i} s_j)(A+B)\right)c(0)\,ds_{i+1} \ldots ds_1.$$

By shifting $0 \to t^n$ and $\tau_n \to t^{n+1}$, we obtain our result:

$$\begin{aligned}
||e_i|| &\leq ||\exp((A+B)\tau_n)c(t^n) - c_i|| \qquad (3.37)\\
&\leq \tilde{C}\tau_n^{i\alpha}||c(t^n)||,
\end{aligned}$$

where $0 \leq \alpha_i < 1$ and i is the number of iteration steps over the operator A, $\tilde{C}$ is a constant and depending only on κ_5 and ω.

$\square$

In the next subsubsection we describe the stability analysis.

3.1.2.2   Stability Analysis

For stability bound we have the following theorem:

Theorem 3.6. *Let us consider the abstract Cauchy problem in a Banach space* **X**

$$||S_i|| \leq \exp(c\omega\tau_n), \qquad (3.38)$$

where c depends only on the coefficients of the method and ω is a bound for the operators, see (3.13)–(3.16). S_i is given as in Equation (3.5) and τ_n is the time-step size.

Proof. We apply the assumption:

$$B = A^{1-\alpha}. \qquad (3.39)$$

Based on the definition of S_i we have:

$$S_i = \exp(A\tau_n) \qquad (3.40)$$

$$+ \int_0^{\tau_n} \exp(As)B \exp((\tau_n - s)A)\, ds$$

$$+ \int_0^{\tau_n} \exp(As_1)B \int_0^{\tau_n - s_1} \exp(s_2 A)B \exp((\tau_n - s_1 - s_2)A)\, ds_2\, ds_1$$

$$+ \ldots +$$

$$+ \int_0^{\tau_n} \exp(As_1)B \int_0^{\tau_n - s_1} \exp(s_2 A)B \int_0^{\tau_n - s_1 - s_2} \exp(s_3 A)B \ldots$$

$$\int_0^{\tau_n - \sum_{j=1}^{i-1} s_j} \exp(As_i)B \exp((\tau_n - \sum_{j=1}^{i-1} s_j)A)\, ds_i \ldots ds_1.$$

After application of $B = A^{1-\alpha}$ we have:

$$\|S_i\| \leq \|\exp(A\tau_n)\| \qquad (3.41)$$

$$+ \left\| \int_0^{\tau_n} \exp(As)A^{1-\alpha} \exp((\tau_n - s)A)\, ds \right\|$$

$$+ \left\| \int_0^{\tau_n} \exp(As_1)A^{1-\alpha} \right.$$

$$\left. \cdot \int_0^{\tau_n - s_1} \exp(s_2 A)A^{1-\alpha} \exp((\tau_n - s_1 - s_2)A)\, ds_2\, ds_1 \right\|$$

$$+ \ldots +$$

$$+ \left\| \int_0^{\tau_n} \exp(As_1)A^{1-\alpha} \right.$$

$$\cdot \int_0^{\tau_n - s_1} \exp(s_2 A)A^{1-\alpha} \int_0^{\tau_n - s_1 - s_2} \exp(s_3 A)A^{1-\alpha} \ldots$$

$$\left. \int_0^{\tau_n - \sum_{j=1}^{i-1} s_j} \exp(As_i)A^{1-\alpha} \exp((\tau_n - \sum_{j=1}^{i-1} s_j)A)\, ds_i \ldots ds_1 \right\|,$$

$$\leq \exp(\omega\tau) + \sum_{j=1}^{i} C_j t^{\alpha j} \leq \exp(\tilde{\omega}\tau) \qquad (3.42)$$

where for all $\omega, C_j \leq 0$ for all $j = 1, \ldots, i$ and $\alpha \in (0,1)$, we find $\tilde{\omega} \leq 0$. Therefore, $\|S_i\| \leq \exp(\tilde{\omega}\tau)$ is bounded.

$\square$

3.1.3 Two-Stage Iterative Schemes for Operators Generating an Analytical Semigroup

In the following we discuss the consistency of the two-stage iterative method, taken into account to iterate over both operators. For such schemes we assume that both operators can generate an analytical semigroup.

Theorem 3.7. *For the numerical solution of (2.40), we consider the iterative operator splitting scheme given with Equations (3.7) and (3.9), meaning we iterate on operator A with $m + 1$ iterative steps and on operator B with m iterative steps.*

If the Assumptions 3.1 and 3.2 are valid, then

$$||S_i^n - \exp((A + B)n\tau)|| \leq C\tau^{(m+1)\alpha - 1}, \ n\tau \leq T, \tag{3.43}$$

where $i = 2m + 1$ are the iterative steps over A and B, the constant C can be chosen uniformly on bounded time intervals and in particular, independent of n and τ. $\alpha \in (0, 1)$ with the assumption $B = A^{1-\alpha}$.

Proof. By applying the telescopic identity we obtain

$$(S_i^n - \exp((A + B)n\tau)u_0 \tag{3.44}$$
$$= \sum_{v=0}^{n-1} S_i^{n-v-1}(S - \exp((A + B)\tau)) \exp(v\tau(A + B))u_0,$$

if we assume the stability bound:

$$||S_i|| \leq \exp(c\omega\tau), \tag{3.45}$$

with a constant c only depends on the estimation of the method.

Furthermore, if we assume the consistency bound:

$$||S_i^n - \exp((A + B)n\tau)|| \tag{3.46}$$
$$\leq \exp(c\omega T) \sum_{v=0}^{n-1} ||(S - \exp(\tau(A + B))) \exp(v\tau(A + B))||$$
$$\leq C\tau^{(m+1)\alpha - 1}, \ n\tau \leq T.$$

The desired consistency and stability bound is given in the next subsubsections.

$\square$

3.1.3.1   Consistency analysis

We present the results of the consistency of our iterative method. We assume for the operator system that the generators of an analytical semigroup are based on their underlying norms for the Banach space $\mathbf{X}$ and that the induced operator norms are denoted by $|| \cdot ||$.

In the following we discuss the consistency of the two-stage iterative method, taken into account to iterate over both operators.

Theorem 3.8. *Let us consider the abstract Cauchy problem in a Banach space $\mathbf{X}$ is given in Equation (3.46). With the operators $A, B : \mathbf{X} \rightarrow \mathbf{X}$ are*

linear operators that are generators of the analytical semigroups. We assume
$dom(B) \subset dom(A)$, *so we are restricted to balance the operators. We assume*

$$B = A^{1-\alpha} \qquad (3.47)$$

is the infinitesimal generator of an analytical semigroup for all $\alpha \in (0,1)$, *see*
[180].

The consistency error is given as $\mathcal{O}(\tau_n^{(m+1)\alpha})$, where $\tau_n = t^{n+1} - t^n$ and
we have equidistant time steps, with $n = 1, \ldots, N$. *Further,* $m + 1$ *are the*
iterative steps with operator A.

Then the iteration process (3.5) for $i = 2m + 1$ for $m = 0, 1, 2, \ldots$, where
we assume $m + 1$ *iterative steps with operator* A *and* m *iterative steps with*
operator B, *is consistent with the order of the consistency* $\mathcal{O}(\tau_n^{\alpha(m+1)})$, *where*
$0 \leq \alpha < 1$.

Proof. Let us consider the iterations (3.7) and (3.9) on the subinterval $[t^n, t^{n+1}]$.
For the first iterations we have:

$$\partial_t c_1(t) = A c_1(t), \quad t \in (t^n, t^{n+1}], \qquad (3.48)$$

and for the second iteration we have:

$$\partial_t c_2(t) = A c_1(t) + B c_2(t), \quad t \in (t^n, t^{n+1}], \qquad (3.49)$$

In general we have:
For the odd iterations for $i = 1, 3, 5, \ldots, 2m + 1$ for $m = 0, 1, 2, \ldots$

$$\partial_t c_i(t) = A c_i(t) + B c_{i-1}(t), \quad t \in (t^n, t^{n+1}], \qquad (3.50)$$

where for $c_0(t) \equiv 0$.
and for the even iterations for $i = 2, 4, 6, \ldots, 2m$ for $m = 1, 2, \ldots$

$$\partial_t c_i(t) = A c_i(t) + B c_{i-1}(t), \quad t \in (t^n, t^{n+1}], \qquad (3.51)$$

where for $c_0(t) \equiv 0$.
This means we iterate at least $m + 1$ times over A and m times over B.
The solutions for the first two iterative steps are given by the variation of
constants:

$$
\begin{aligned}
c_1(t) &= \exp(A(t - t^n))c(t^n), \quad t \in (t^n, t^{n+1}], \qquad &(3.52)\\
c_2(t) &= \exp(B(t - t^n))c(t^n) \qquad &(3.53)
\end{aligned}
$$

$$+ \int_{t^n}^{t^{n+1}} \exp(B(t^{n+1} - s))A c_1(s) ds, \quad t \in (t^n, t^{n+1}].$$

For the odd iterations $i = 2m + 1$ for $m = 0, 1, 2, \ldots$ we obtain the

$$
\begin{aligned}
c_i(t) &= S_i(t)c(t^n) \qquad &(3.54)\\
&= \exp(A(t - t^n))c(t^n)
\end{aligned}
$$

$$+ \int_{t^n}^{t} \exp((t - s)A)B c_{i-1}(s) \, ds, \quad t \in (t^n, t^{n+1}].$$

For the even iterations $i = 2m$ for $m = 1, 2, \ldots$ we obtain the

$$
\begin{aligned}
c_i(t) &= S_i(t)c(t^n) \\
&= \exp(B(t - t^n))c(t^n) \\
&\quad + \int_{t^n}^t \exp((t - s)B)Ac_{i-1}(s)\, ds, \quad t \in (t^n, t^{n+1}].
\end{aligned}
\tag{3.55}
$$

The consistency is given as in the following steps:

For e_1 we have the result of the previous one-stage iterative operator method:

$$
\|e_1\| \leq C\tau^\alpha \|c(t^n)\|
\tag{3.56}
$$

where $\alpha \in (0, 1)$, $\tau = (t^{n+1} - t^n)$ and C is a constant depends only on κ_5 and ω.

For e_2 we have to apply the even steps:

$$
\begin{aligned}
c_2(t^{n+1}) &= \exp(B\tau_n)c(t^n) \\
&\quad + \int_{t^n}^{t^{n+1}} \exp(B(t^{n+1} - s))A \exp((s - t^n)B)c(t^n)\, ds,
\end{aligned}
\tag{3.57}
$$

$$
\begin{aligned}
c(t^{n+1}) &= \exp(B\tau_n)c(t^n) \\
&\quad + \int_{t^n}^{t^{n+1}} \exp(B(t^{n+1} - s))A \exp((s - t^n)B)c(t^n)\, ds \\
&\quad + \int_{t^n}^{t^{n+1}} \exp(B(t^{n+1} - s))A \\
&\qquad \int_{t^n}^s \exp(B(s - \rho))A \exp((\rho - t^n)(A + B))c(t^n)\, d\rho\, ds.
\end{aligned}
\tag{3.58}
$$

We obtain:

$$
\|e_2\| \leq \| \exp((A + B)\tau_n)c(t^n) - c_2 \|
\tag{3.59}
$$

$$
= \| \int_{t^n}^{t^{n+1}} \exp(B(t^{n+1} - s))A
\tag{3.60}
$$

$$
\int_{t^n}^s \exp(B(s - \rho))A \exp((\rho - t^n)(A + B))c(t^n)\, d\rho\, ds \|
$$

$$
= \int_{t^n}^{t^{n+1}} \| \exp(B(t^{n+1} - s)) \|
\tag{3.61}
$$

$$
\int_{t^n}^s \| \exp(B(s - \rho))A^{2-\alpha} \exp((\rho - t^n)(A + B))c(t^n)\, d\rho \| ds
$$

$$
= \int_{t^n}^{t^{n+1}} C \int_{t^n}^s (s - \rho)^{\alpha-2} d\rho ds \|c(t^n)\|
\tag{3.62}
$$

$$
\leq C\tau^\alpha \|c(t^n)\|
$$

where $\alpha \in (0,1)$, $\tau = t^{n+1} - t^n$ and C is a constant only depending on κ_2, κ_5 and ω.

For the general iterative steps, the recursive proof is given in the following. We shift $t^n \to 0$ and $t^{n+1} \to \tau_n$ for simpler calculations, see [136]. The initial conditions are given with $c(0) = c(t^n)$.

For the odd iterative steps $i = 2m + 1$ with $m = 0, 1, 2, \ldots$, we have the result of the previous one-stage iterative operator method:

$$||e_i|| \leq \tilde{C} \tau_n^{(m+1)\alpha} ||c(t^n)||,$$

where $0 \leq \alpha_i < 1$ and $m+1$ are the number of iteration steps over the operator A, $\tilde{C}$ is a constant and depending only on κ_2, κ_5, and ω.

For the even iterative steps $i = 2m + 1$ with $m = 1, 2, \ldots$, we have m iterative steps with A and m iterative steps with B. We obtain for c_i and c:

$$c_i(\tau_n) = \exp(B\tau_n)c(0) \tag{3.63}$$
$$+ \int_0^{\tau_n} \exp(Bs)A\exp((\tau_n - s)A)c(0)\,ds$$
$$+ \int_0^{\tau_n} \exp(Bs_1)A \int_0^{\tau_n - s_1} \exp(s_2 A)B\exp((\tau_n - s_1 - s_2)B)c(0)\,ds_2\,ds_1$$
$$+ \ldots +$$
$$+ \int_0^{\tau_n} \exp(Bs_1)A \int_0^{\tau_n - s_1} \exp(s_2 A)B \int_0^{\tau_n - s_1 - s_2} \exp(s_3 B)A\ldots$$
$$\int_0^{\tau_n - \sum_{j=1}^{i-1} s_j} \exp(Bs_i)A\exp\left((\tau_n - \sum_{j=1}^{i-1} s_j)A\right)c(0)\,ds_i \ldots ds_1,$$

$$c(\tau_n) = \exp(B\tau_n)c(0) \tag{3.64}$$
$$+ \int_0^{\tau_n} \exp(Bs)A\exp((\tau_n - s)A)c(0)\,ds$$
$$+ \int_0^{\tau_n} \exp(Bs_1)A \int_0^{\tau_n - s_1} \exp(s_2 A)B\exp((\tau_n - s_1 - s_2)B)c(0)\,ds_2\,ds_1$$
$$+ \ldots +$$
$$+ \int_0^{\tau_n} \exp(Bs_1)A \int_0^{\tau_n - s_1} \exp(s_2 A)B \int_0^{\tau_n - s_1 - s_2} \exp(s_3 B)A\ldots$$
$$\int_0^{\tau_n - \sum_{j=1}^{i-1} s_j} \exp(Bs_i)A\exp\left((\tau_n - \sum_{j=1}^{i-1} s_j)A\right)c(0)\,ds_i \ldots ds_1$$
$$+ \int_0^{\tau_n} \exp(As_1)B \int_0^{\tau_n - s_1} \exp(s_2 A)B \int_0^{\tau_n - s_1 - s_2} \exp(s_3 A)B\ldots$$
$$\int_0^{\tau_n - \sum_{j=1}^{i} s_j} \exp(As_{i+1})B\exp\left((\tau_n - \sum_{j=1}^{i} s_j)(A + B)\right)c(0)\,ds_{i+1} \ldots ds_1.$$

By shifting $0 \to t^n$ and $\tau_n \to t^{n+1}$, we obtain our result:

$$\|e_i\| \leq \| \exp((A + B)\tau_n)c(t^n) - c_i\| \tag{3.65}$$
$$\leq \tilde{C}\tau_n^{m\alpha}\|c(t^n)\|,$$

where $0 \leq \alpha_i < 1$ and i is the number of iteration steps over the operator A, $\tilde{C}$ is a constant and depending only on κ_2, κ_5, and ω.

$\square$

In the next subsubsection we describe the stability analysis.

3.1.3.2   Stability Analysis

For stability bound we have the following theorem:

Theorem 3.9. *Let us consider the abstract Cauchy problem in a Banach space* **X**

$$\|S_i\| \leq \exp(c\omega\tau_n), \tag{3.66}$$

where c depends only on the coefficients of the method and ω is a bound for the operators, see Assumptions (3.13)–(3.16). S_i is given as in Equation (3.54) and τ_n is the time-step size.

Proof. We apply the assumption:

$$B = A^{1-\alpha}. \tag{3.67}$$

Based on the definition of S_i we have: For the even iterative steps $i = 2m+1$ with $m = 1, 2, \ldots$, we have m iterative steps with A and m iterative steps with B. We obtain for c_i and c:

$$S_i = \exp(B\tau_n) \tag{3.68}$$
$$+ \int_0^{\tau_n} \exp(Bs)A\exp((\tau_n - s)A)\,ds$$
$$+ \int_0^{\tau_n} \exp(Bs_1)A\int_0^{\tau_n-s_1} \exp(s_2A)B\exp((\tau_n - s_1 - s_2)B)\,ds_2\,ds_1$$
$$+ \ldots +$$
$$+ \int_0^{\tau_n} \exp(Bs_1)A\int_0^{\tau_n-s_1} \exp(s_2A)B\int_0^{\tau_n-s_1-s_2} \exp(s_3B)A\ldots$$
$$\int_0^{\tau_n-\sum_{j=1}^{i-1} s_j} \exp(Bs_i)A\exp((\tau_n - \sum_{j=1}^{i-1} s_j)A)\,ds_i\ldots ds_1,$$

After application of B we have:

$$||S_i|| = ||\exp(B\tau_n)|| \tag{3.69}$$

$$+||\int_0^{\tau_n} \exp(Bs)A\exp((\tau_n - s)A)\,ds||$$

$$+||\int_0^{\tau_n} \exp(Bs_1)A\int_0^{\tau_n-s_1} \exp(s_2A)B\exp((\tau_n - s_1 - s_2)B)\,ds_2\,ds_1||$$

$$+\ldots+$$

$$+||\int_0^{\tau_n} \exp(Bs_1)A\int_0^{\tau_n-s_1} \exp(s_2A)B\int_0^{\tau_n-s_1-s_2} \exp(s_3B)A\ldots$$

$$\int_0^{\tau_n-\sum_{j=1}^{i-1}s_j} \exp(Bs_i)A\exp((\tau_n - \sum_{j=1}^{i-1}s_j)A)\,ds_i\ldots ds_1||,$$

$$\leq \exp(\omega\tau) + \sum_{j=1}^{i-1} C_j t^{\alpha j} \leq \exp(\tilde{\omega}\tau) \tag{3.70}$$

where for all $\omega, C_j \leq 0$ for all $j = 1,\ldots,i-1$ and $\alpha \in (0,1)$, we find $\tilde{\omega} \leq 0$.
Therefore, $||S_i|| \leq \exp(\tilde{\omega}\tau)$ is bounded.
The same can be done for the odd iterations.

$\square$

3.1.4   Some Examples for One-Stage and Two-Stage Iterative Operator Splitting Schemes

In the following, we describe some examples for one-stage and two-stage iterative operator splitting schemes.

3.1.4.1   One-Stage Iterative Scheme

1. Example convection-diffusion equation
We assume:

$A = \nabla D\nabla$, and

$B = -\mathbf{v}\cdot\nabla$, where $D \in \mathbb{R}^+$ and $\mathbf{v} \in \mathbb{R}^m$ with m is the dimension.

We assume $\alpha = \frac{1}{2}$ and apply two iterative steps over operator A and have the following local errors:

$$||e_1|| = \tilde{C}\tau_n^{\frac{1}{2}} \tag{3.71}$$

and hence,

$$||e_2|| = \tilde{\tilde{C}}\tau_n^1, \tag{3.72}$$

where $\tilde{C}, \tilde{\tilde{C}}$ are constants independent of τ_n.

2. Example diffusion-reaction equation
We assume:
$Au = \nabla D \nabla u$ and

$B = -\lambda u$, where $D \in \mathbb{R}^+$ and $\lambda \in \mathbb{R}$ is the reaction part.

We assume $\alpha = 0$ and apply two iterative steps over operator A and have the following local errors:

$$||e_1|| = \tilde{C} \tau_n^1 \tag{3.73}$$

and hence,

$$||e_2|| = \tilde{\tilde{C}} \tau_n^2, \tag{3.74}$$

where $\tilde{C}, \tilde{\tilde{C}}$ are constants independent of τ_n.

Remark 3.10. Here we have the full convergence order of the method, while the B operator is unconditionally smooth.

3. Example diffusion-reaction equation with spatial dependent terms:
The differential equation

$$\frac{\partial c}{\partial t} = \partial_{xx} u - x^2 u + \sqrt{1 + x^2} u, \tag{3.75}$$

$$\frac{\partial c}{\partial t} = Au + Bu, \tag{3.76}$$

We assume: $A : D(A) \to \mathbf{X}$, $\mathbf{X} \in L_2(\mathbb{R})$, $B : \mathbf{X} \to \mathbf{X}$.

$Au = \partial_{xx}$ and

$B = \sqrt{1 + x^2}$,

where $D(A) = \{u \in H^2(\mathbb{R}) | x^2 u(x) \in L_2(\mathbb{R})\}$.

We can bound:

$$||Bu|| = C(||A^\alpha u|| + ||u||)$$

and $\alpha = 1/2$.

We apply two iterative steps over operator A and have the following local errors:

$$||e_1|| = \tilde{C} \tau_n^{1/2} \tag{3.77}$$

and hence,

$$||e_2|| = \tilde{\tilde{C}} \tau_n^1, \tag{3.78}$$

where $\tilde{C}, \tilde{\tilde{C}}$ are constants independent of τ_n.

Remark 3.11. Here we have the full convergence order of the method, while the B operator is unconditionally smooth with respect to the spatial derivation.

3.1.4.2   Two-Stage Iterative Scheme

1. Example biharmonic equation
The differential equation

$$\frac{\partial c}{\partial t} = -\Delta^2 u + u\Delta u, \tag{3.79}$$

$$\frac{\partial c}{\partial t} = Au + Bu, \tag{3.80}$$

We assume:

$$Au = -\Delta^2,$$

$$B = (-A)^{\frac{1}{2}}.$$

We have $\alpha = \frac{1}{2}$ and apply two-stages of iterative steps over operator A and one over operator B. We have the following local errors:

$$||e_1|| = \tilde{C}\tau_n^{\frac{1}{2}} \tag{3.81}$$

and hence

$$||e_2|| = \tilde{\tilde{C}}\tau_n^{\frac{1}{2}}, \tag{3.82}$$

where $\tilde{C}, \tilde{\tilde{C}}$ are constants independent of τ_n.

Remark 3.12. Here we have at least only an improvement, when we iterate over operator A. By the way, in the numerical experiments, we also claim for such two-stage schemes higher-order results.

2.) Example Heat-equation
The differential equation

$$\frac{\partial c}{\partial t} = -\nabla \cdot D_1 \nabla u - \nabla \cdot D_2 \nabla u, \tag{3.83}$$

$$\frac{\partial c}{\partial t} = Au + Bu, \tag{3.84}$$

where $D_1, D_2 \in \mathbb{R}^{m \times m}$ and m is the dimension.
We assume:

$$A = -\nabla \cdot D_1 \nabla,$$

$$B = -\nabla \cdot D_2 \nabla,$$

We have $\alpha = 0$ and apply two-stages of iterative steps over operator A and one over operator B. We have the following local errors:

$$||e_1|| = \tilde{C}\tau_n^0 \qquad (3.85)$$

and hence

$$||e_2|| = \tilde{\tilde{C}}\tau_n^0, \qquad (3.86)$$

where $\tilde{C}, \tilde{\tilde{C}}$ are constants independent of τ_n.

Remark 3.13. Here we have the worst case. We cannot regain some smoothness of the other operator and we have lost at least the convergence order. By the way, experiments show that we also have higher-order results, meaning that one operator is at least smoother, e.g., $D_1 \gg D_2$, so D_2 smooths the scheme.

Chapter 4

Computation of the Iterative Splitting Methods: Algorithmic Part

In the following, we discuss standard schemes based on Runge-Kutta (RK) methods to compute iterative schemes and novel methods based on matrix exponentials to compute iterative schemes in a fast way. We have also discussed the ideas and theoretical results more detailed in [112].

4.1 Exponential Runge-Kutta Methods to Compute Iterative Splitting Schemes

In the following, we concentrate on applying a exponential Runge-Kutta method to iterative splitting schemes.

Here, the idea is to take into account the exponential behavior of the equations with a right-hand side. Such specific Runge-Kutta methods are developed in such a direction to apply inhomogeneous exponential-dependent functions.

We discuss an exponential Runge-Kutta method that is given in [132].

The underlying ODE system is given as:

$$\frac{dy}{dt} = Ay + f(y,t), \ t \in [0,T], \tag{4.1}$$

$$y(0) = y_0, \tag{4.2}$$

where $y : \mathbb{R} \to \mathbb{R}^n$ is the solution function and the initial condition $y_0 \in \mathbb{R}^n$ is given.

Based on the equation and with the variation of constants, we obtain the exponential function.

A simple exponential Runge-Kutta method is given as:

$$
\begin{array}{c|cc}
0 & & \\
c_2 & a_{21} = b_2\phi_{1,2} & \\
\hline
& b_1 = (1 - \frac{1}{2c_2})\phi_1 & b_2 = \frac{1}{c_2}\phi_1
\end{array}
\tag{4.3}
$$

where for $b_2 = \frac{1}{2}$ and

$$\phi_0(A\tau) = \exp(A\tau), \tag{4.4}$$

$$\phi_1(A\tau) = \frac{\exp(A\tau) - I}{A\tau}, \tag{4.5}$$

$$\phi_2(A\tau) = \frac{\phi_1(A\tau) - I}{A\tau}, \tag{4.6}$$

and

$$\phi_{i,j} = \phi_{i,j}(A\tau) = \phi_i(c_j A\tau), \ 2 \leq j \leq s, \tag{4.7}$$

and s is the stage of the RK method.
The ϕ-function is given as

$$\phi_0(x) = e^x, \tag{4.8}$$

$$\phi_{k+1}(x) = \frac{\phi_k(x) - \frac{1}{k!}}{x}, \ k \geq 0.$$

We start to derive the iterative scheme with respect to the application of the exponential Runge-Kutta method.

$$\frac{\partial y_1}{\partial t} = Ay_1 + By_0, \tag{4.9}$$

$$\frac{\partial y_2}{\partial t} = Ay_1 + By_2, \tag{4.10}$$

$$\cdots \tag{4.11}$$

where $y_0(t) = \exp(At)\exp(Bt)y(t^n)$.

The iterated scheme is given as:

$$y_1(t) = \exp(At)y(t^n) + \int_0^t \exp(A(t-s))By_0(s)\,ds, \tag{4.12}$$

and we obtain in the $s = 2$ Runge-Kutta stage notation:

$$y_1(t) = \phi_0(At)y(t^n) + t\,b_1(At)B\,y_0(t^n) + t\,b_2(At)B\,y_0(t^n + c_2t), \tag{4.13}$$

and we have for $y_2(t)$:

$$y_2(t) = \phi_0(Bt)y(t^n) + \int_0^t \exp(B(t-s))Ay_1(s)\,ds \tag{4.14}$$

and we obtain in the $s = 3$ Runge-Kutta stage notation:

$$\begin{aligned} y_2(t) = {} & \phi_0(Bt)y(t^n) + t\,b_1(Bt)A\,y_1(t^n) + t\,b_2(Bt)A\,y_0(t^n + c_2t) \\ & + \ t\,b_3(Bt)B\,y_0(t^n + c_3t). \end{aligned} \tag{4.15}$$

Remark 4.1. The iterative schemes can be computed by the exponential Runge-Kutta schemes. By the way, higher Runge-Kutta schemes are necessary for higher-order iterative schemes, e.g., $i = 2$ needs a third-order exponential Runge-Kutta scheme. For practical application, we are restricted to second- and third-order schemes. We also have to compute the intermediate time steps that are time-consuming.

4.2   Matrix Exponentials to Compute Iterative Splitting Schemes

The theoretical ideas can be discussed in the following formulation:

$$D_A(B,t)^{[0]} = \exp(tB), \tag{4.16}$$

$$D_A(B,t)^{[k]} = k \int_0^t \exp((t-s)B)\, A\, D_A(B,s)^{[k-1]} ds$$

and the matrix formulation of our two-step scheme is given as:

$$\tilde{A} = \begin{pmatrix} A & 0 & \cdots & \cdots \\ A & B & 0 & \cdots \\ 0 & B & A & \cdots \\ \vdots & \ddots & \ddots & \ddots \\ 0 & \cdots & B & A \end{pmatrix}, \tag{4.17}$$

the computation of the exp-Matrix can be expressed as:

$$\exp(\tilde{A}t) := \begin{pmatrix} \exp(At) & 0 & \cdots & \cdots \\ D_A(B) & \exp(Bt) & 0 & \cdots \\ D_B(A)^{[2]}/2! & D_B(A)^{[1]}/1! & \exp(At) & \cdots \\ \vdots & \ddots & \ddots & \ddots \\ D_B(A)^{[n]}/n! & \cdots & D_B(A)^{[1]}/1! & \exp(At) \end{pmatrix}. \tag{4.18}$$

Here we have to compute the right-hand side as a time-dependent term, which means we evaluate $\exp(At)$ and $\exp(Bt)$ as a Taylor expansion. Then the integral formulation can be done with polynomials and we could derive ϕ-functions.

In the following, we reduce to an approximation of the fixed right-hand side (meaning we assume $D_A(B,s)^{[k-1]} \approx D_A(B,0)^{[k-01]}$).

Later we also follow with more extended schemes.

Computing ϕ-functions:

$$\phi_0(x) \;=\; e^x, \tag{4.19}$$

$$\phi_{k+1}(x) \;=\; \frac{\phi_k(x) - \frac{1}{k!}}{x}, \quad k \geq 0.$$

So the matrix formulation of our scheme is given as

$$y(t) = \exp(\tilde{A}t)y(0) \,. \tag{4.20}$$

where y is a solution function in $\mathbb{R}^n$ and $\tilde{A}$ is given as,

$$\tilde{A} = \begin{pmatrix} A & 0 & \cdots & \cdots \\ A & B & 0 & \cdots \\ 0 & A & B & \cdots \\ \vdots & \ddots & \ddots & \ddots \\ 0 & \cdots & B & A \end{pmatrix}, \tag{4.21}$$

the computation of the exp-Matrix can be expressed in a first-order scheme and with the assumption of commutations is given as:

$$\tag{4.22}$$

$$\exp(\tilde{A}t) = \begin{pmatrix} \exp(At) & 0 & \cdots & \cdots & \cdots \\ \phi_1(Bt)A\,t & \exp(Bt) & 0 & \cdots & \cdots \\ \phi_2(At)B(B+A)\,t^2 & \phi_1(At)B\,t & \exp(At) & \cdots & \cdots \\ \vdots & & \ddots & \ddots & \vdots \\ \phi_i(At)B(A+B)^{i-1}\,t^i & \cdots & & \phi_1(At)B\,t & \exp(At) \end{pmatrix},$$

where we assume $u_i(s) = u(t^n)$, meaning we approximate the right-hand side term.

For higher orders we should also include the full derivations of $c_1, c_2, \ldots$.

4.2.1   Derivation of the Formula

Consider the equation

$$\dot{u} = Au + a, \quad u(0) = 0, \tag{4.23}$$

the solution of this equation in terms of the $\phi_k, k = 0, 1$, is given as

$$u = t\,\phi_1(tA)a. \tag{4.24}$$

Similarly the equation

$$\dot{u} = Au + bt, \quad u(0) = 0, \tag{4.25}$$

has a solution of the form

$$u = t^2\,\phi_2(tA)b. \tag{4.26}$$

In general, the solution of the equation

$$\dot{u} = Au + a + bt + c\frac{t^2}{2!} + ..., \quad u(0) = 0, \tag{4.27}$$

admits the form

$$u = t\,\phi_1(tA)a + t^2\,\phi_2(tA)b + t^3\,\phi_3(tA)c + ... \tag{4.28}$$

In this section, we use the formula (4.42) for iterative schemes given as

$$\begin{align}
\dot{u}_1 &= Au_1, \tag{4.29}\\
\dot{u}_2 &= Au_1 + Bu_2, \tag{4.30}\\
\dot{u}_3 &= Au_3 + Bu_2, \tag{4.31}\\
\dot{u}_4 &= ... \tag{4.32}
\end{align}$$

The solution of the first iteration is given by

$$u_1 = e^{At}u_0, \tag{4.33}$$

Inserting this into Equation (5.24) and expanding e^{At} up to the first order, we have second order approximation of the exact solution that has a form

$$u_2 = e^{Bt}u_0 + \phi_1(tA)A\,tu_0. \tag{4.34}$$

Similarly, inserting (4.34) into the Equation (4.50) and expanding $\phi_1(tA)$, we have a third-order approximation of the exact solution

$$u_3 = e^{Bt}u_0 + \phi_1(tB)B\,tu_0 + \phi_2(tB)B(B+A)\,t^2u_0. \tag{4.35}$$

In general for $i = 0, 2, 4, ...$, we have the $p = i + 1$-th order approximation of the exact solution in terms of the ϕ_i function as follows

$$\begin{align}
u_i = \;& e^{At}u_0 + \phi_1(tA)A\,tu_0 + \phi_2(tA)A + (B+A)\,t^2u_0\\
& ... + \phi_i(tA)A(A+B)^{i-1}u_0. \tag{4.36}
\end{align}$$

For $i = 1, 3, 5, ...$, we have

$$\begin{align}
u_i = \;& e^{Bt}u_0 + \phi_1(tB)B\,tu_0 + \phi_2(tB)B + (B+A)\,t^2u_0\\
& ... + \phi_i(tB)B(B+A)^{i-1}\,t^iu_0. \tag{4.37}
\end{align}$$

4.3   Algorithms

The algorithmic ideas of the two- and one-side iterative schemes are discussed in the following subsections.

4.3.1   Two-Side Scheme

For the implementation of the integral formulation, we have to deal with parallel ideas, meaning we select independent parts of the formulation.

- Step 1: Determine the order of the method by fixed iteration number.

- Step 2: Consider the time interval $[t_0, T]$, divide it into N subintervals so that the time step is $h = (T - t_0)/N$.

- Step 3: On each subinterval, $[t_n, t_n+h]$, $n = 0, 1, ..., N$, use the algorithm by considering initial conditions for each step as $u(t_0) = u_0$, $u_i(t_n) = u_{i-1}(t_n) = u(t_n)$,

$$
\begin{align}
u_2(t_n + h) &= \left(\phi_0(Bt) + \phi_1(Bt)A\, t\right) u(t_n) \tag{4.38}\\
u_3(t_n + h) &= \left(\phi_0(At) + \phi_1(At)B\, t\right.\notag\\
&\quad \left. + \phi_2(At)B(B + A)\, t^2\right) u(t_n) \tag{4.39}
\end{align}
$$

$$\vdots$$

$$
\begin{align}
u_{2i}(t_n + h) &= \left(\phi_0(Bt) + \phi_1(Bt)A\, t + \right. \tag{4.40}\\
&\quad \left. \ldots + \phi_{2i-1}(Bt)A(A + B)^{2i-1}\, t^{2i}\right) u_{2i-1}(t_n)\notag\\
u_{2i+1}(t_n + h) &= \left(\phi_0(At) + \phi_1(At)B\, t + \right. \tag{4.41}\\
&\quad \left. \ldots + \phi_{2i}(At)B(B + A)^{2i}\, t^{2i+1}\right) u_{2i}(t_n)\notag
\end{align}
$$

- Step 4: $u_i(t_n + h) \rightarrow u(t_n + h)$

- Step 5: Repeat this procedure for the next interval until the desired time T is reached.

4.3.2   One-Side Scheme (Alternative Notation with Commutators)

For the one-side scheme, we taken into account of the following commutator relation

Definition 4.2. *The following relation is given:*

$$\exp(-tA)B\exp(tA) = [B, \exp(tA)] \tag{4.42}$$

where $[\cdot, \cdot]$ is the matrix commutator.
The integration is given as:

$$\int_0^t \exp(-sA)B\exp(sA)ds = [B, \phi_1(tA)] \tag{4.43}$$

the representation is given in [165].

Further, we have the recursive integration:

$$[B, \phi_i(tA)t^i] = \int_0^t [B, \phi_{i-1}(sA)] \, ds \qquad (4.44)$$

where ϕ_i is given as:

$$\phi_0(At) = \exp(At), \qquad (4.45)$$

$$\phi_i(At) = \int_0^t \phi_{i-1}(As) \, ds, \qquad (4.46)$$

$$\phi_i(At) = \frac{\phi_{i-1}(At) - I \frac{t^{i-1}}{(i-1)!}}{A}. \qquad (4.47)$$

The iterative scheme with the equations:

$$\dot{u}_1 = Au_1, \qquad (4.48)$$
$$\dot{u}_2 = Au_2 + Bu_1, \qquad (4.49)$$
$$\dot{u}_3 = Au_3 + Bu_2, \qquad (4.50)$$
$$\dot{u}_4 = \ldots \qquad (4.51)$$

is solved as:

$$c_1(t) = \exp(At)c(t^n), \qquad (4.52)$$

$$c_2(t) = c_1(t) + c_1(t) \int_0^t [B, \exp(sA)] ds \,,$$

$$c_2(t) = c_1(t) + c_1(t) [B, \phi_1(tA)], \qquad (4.53)$$

$$c_3(t) = c_2(t) + c_1(t) \int_0^t [B, \exp(sA)][B, \phi_1(sA)] ds \,, \qquad (4.54)$$

$$c_3(t) = c_2(t) + c_1(t) \left([B, \exp(tA)][B, \phi_2(tA)] \right. \qquad (4.55)$$

$$\left. + [B, A \exp(tA)][B, \phi_3(tA)] \right) + O(t^3) \,,$$

$$\ldots$$

The recursion is given as:

$$c_i(t) = c_{i-1}(t) \qquad (4.56)$$

$$+ c_1(t) \int_0^t [B, \exp(sA)]$$

$$\cdot \int_0^{s_1} [B, \exp(s_1 A)] \int_0^{s_2} \ldots \int_0^{s_{i-2}} [B, \exp(s_{i-1} A)] ds_{i-1} \ldots ds_1 dt.$$

Remark 4.3. For the novel notation, we have embedded the commutator to the computational scheme. For such a scheme we could save to compute the additional commutators.

Chapter 5

Extensions of Iterative Splitting Schemes

Iterative splitting schemes can be extended to different applications and therefore various schemes exist that respect the underlying operators A and B, which can be time dependent, spatially discretized, or nonlinear; see the extensions in the works [86], [91], [92], and [93]. There exists no general scheme and all iterative splitting schemes have to be discussed with reference to the special underlying differential equation, e.g., the reaction-diffusion equation [128] and the convection-diffusion equation [95]. In all the extensions that follow, we assume that we have bounded operators, which can be given by spatial discretization or by an underlying real-valued matrix.

Figure 5.1 presents the research directions that are followed in the numerical experiments, see also Chapter 6.

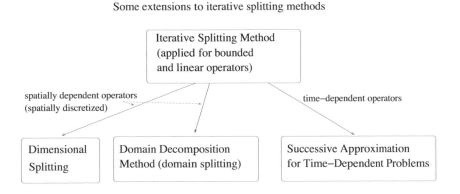

FIGURE 5.1: Some extensions to the iterative splitting schemes.

5.1   Embedded Spatial Discretization Methods to Iterative Splitting Methods

In this section, we discuss the spatial discretization schemes that we apply in splitting methods to semidiscretized convection-diffusion equations. The methods are based on finite difference schemes, see [161].

We propose spatial discretization schemes, which can be modified to higher-order schemes and are embedded to the iterative schemes in Chapter 6.

One important point is that the order of the spatial schemes have to be the same or even a higher-order scheme then the underlying iterative splitting scheme. So it is necessary to have second-order or higher-order spatial schemes to obtain at least with two or more iterative steps a second order scheme for all involved schemes.

Further the spatial and time discretization schemes have to be balanced, such that the resulting order of the splitting scheme is not reduced by a lower discretization order, e.g., an order reduction based on a stiff operator (see [151]).

An important aspect of spatial and time discretization schemes is the balance between the time and spatial schemes, which is discussed in the next subsection.

5.1.1   Balancing of Time and Spatial Discretization

Splitting methods are important for partial differential equations, because they reduce the computational time needed to solve the differential equations and they accelerate the solver process, see [93].

Here, additional balancing is taken into account, because of the spatial step. We assume that the operators are bounded operators, because of the semidiscretization of our differential equations.

The following theorem addresses the delicate situation involving time and spatial steps and the way that the order of the scheme is reduced theoretically.

Remark 5.1. We solve the initial value problem by applying iterative operator splitting schemes to Equations (2.40)–(2.41). We assume bounded and constant operators A, B (e.g., derived by a spatial discretization method, the Method of Lines). When iterating in i-time with A and in j-time with B, the theoretical order is given as $O(\tau_n^{i+j})$.

We obtain a reduction in the order of the iterative scheme to $O(\tau_n^i)$, if the norm of B is equal to or larger than $O(\frac{1}{\tau_n})$. The same reduction can also be obtained with operator A.

So the balancing of time and space is important, see [192].

Remark 5.2. By using an explicit method for the time discretization, we are restricted, for example, by the CFL condition to applying our scheme to the convection-diffusion equation. We have coupled the temporal and spatial scales, see [192], and in the experiments we see that we obtain more accurate results.

On the other hand, by using an implicit method for the time discretization, we do not couple the temporal and spatial scales by a CFL condition and so the scheme remains independent of the reduction but less accurate results are obtained, see Subsection 6.3.2.

In the following section, we discuss such balancing ideas.

5.1.2 Spatial Discretization Schemes with Dimensional Splitting

In the following, we apply our splitting method to partial differential equations, which are spatially discretized. To apply our method to partial differential equations, we have to consider higher-order spatial discretization schemes.

A very promising method is based on Lax-Wendroff schemes for which we achieve second-order results. One of the main problems is to balance time and spatial discretization methods between the advection and diffusion operator. For both we can derive stable schemes, taking into account the stability conditions of each operator; see also [161].

In the following, we discuss an efficient scheme based on the Lax-Wendroff scheme.

5.1.2.1 The Lax-Wendroff Scheme in One Dimension

The advection-diffusion equation in one dimension is:

$$
\begin{aligned}
\partial_t u &= -v\partial_x u + \partial_x D\partial_x u, \ (x,t) \in \Omega \times [0,T], \\
u(x,t_0) &= \tilde{u}_0(x), \ x \in \Omega, \\
u(x,t) &= \tilde{u}_1, \ (x,t) \in \partial\Omega \times [0,T],
\end{aligned}
$$

where $v, D \in \mathbb{R}^+$ and $\Omega \subset \mathbb{R}$, $T \in \mathbb{R}^+$, $\tilde{u}_0 : \Omega \to \mathbb{R}^+$ is the initial function and $\tilde{u}_1 \in \mathbb{R}^+$ is a constant.

The second order Lax-Wendroff scheme for the 1D advection-diffusion equation is:

$$
\begin{aligned}
u_j(t^{n+1}) &= u_j(t^n) + \frac{\mathrm{v}}{2}(u_{j+1}(t^n) - u_{j-1}(t^n)) \\
&\quad - \frac{z\mathrm{v}^2}{2}(u_{j+1}(t^n) - 2u_j(t^n) + u_{j-1}(t^n)),
\end{aligned} \tag{5.1}
$$

where $u_j = u(x_j)$ is the solution at grid point $x_j = j\Delta x$ with $j \in \{0, 1, \ldots, J\}$, $J \in \mathbb{N}^+$ and $\Delta x > 0$ is the spatial step size. The time interval $[0,T]$ is given with the time points $t_n = n\Delta t$, where Δt is the equidistant time step. Further, $\mathrm{v} = \frac{v\Delta t}{\Delta x}$ is the Courant-Friedrich-Levy number and

$$
z(D, v, \Delta t) = \frac{2D}{v^2\Delta t} + 1. \tag{5.2}
$$

Since the Lax-Wendroff scheme is suitable for hyperbolic partial differential equations (hyperbolic PDEs) only, we have to examine the stability closely and carry out a von-Neumann stability analysis.

In the following we sketch the extension ideas with respect to the CFL condition. A more detailed analysis is discussed in Chapter 8.3 in the book [160].

We encounter the following stability condition for the Fourier coefficients of $u_j(t^n)$:

$$\left| \frac{C_k^{n+1}}{C_k^n} \right|^2 = (1 + z\mathbf{v}^2 \cos(\theta_k) - z\mathbf{v}^2)^2 + \mathbf{v}^2 \sin^2(\theta_k) \leq 1 \tag{5.3}$$

$$\Rightarrow \cos(\theta_k) \leq \frac{\sqrt{(1 - z\mathbf{v}^2)^2 + (z^2\mathbf{v}^2 - 1)(2z - z^2 - 1)} + z\mathbf{v}^2 - 1}{z^2\mathbf{v}^2 - 1}. \tag{5.4}$$

Obviously, this statement is always true for right-hand sides ≥ 1, so it is sufficient to estimate the solutions of

$$\Rightarrow z^2\mathbf{v}^2 - z\mathbf{v}^2 \leq \sqrt{(1 - z\mathbf{v}^2)^2 + (z^2\mathbf{v}^2 - 1)(2z - z^2 - 1)}. \tag{5.5}$$

In Figure 5.2, we see the pairs (v, z) for which the statement holds. Of course, only the region $\{(v, z), v \geq 0, z \geq 1\}$ is of practical interest. The figure suggests that, for all $v \geq 0$ (means CFL conditions), there are $z \geq 1$ leading to a stable scheme. For small D for which this is true, we may say that the problem is convection dominated and for large diffusion, the CFL condition is nearly zero to stabilize the scheme.

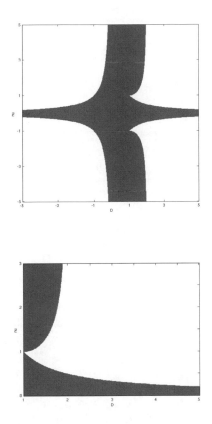

FIGURE 5.2: Stability condition of the Lax-Wendroff scheme: (v, z) inside the blue area satisfy the stability condition. The region of interest is also shown in detail, where the x-axis is v (CFL-condition) and the y-axis is D (diffusion).

In the next subsection, we discuss the generalization to two dimensions.

5.1.2.2   Generalization to Two Dimensions

The advection-diffusion equation in two dimensions is:

$$
\begin{aligned}
\partial_t u &= -\mathbf{v} \cdot \nabla u + \nabla \cdot D \nabla u, \ (\mathbf{x}, t) \in \Omega \times [0, T] \\
&= -v_x \frac{\partial u}{\partial x} - v_y \frac{\partial u}{\partial y} + D \frac{\partial^2 u}{\partial x^2} + D \frac{\partial^2 u}{\partial y^2}, \\
u(\mathbf{x}, t_0) &= \tilde{u}_0(\mathbf{x}), \ \mathbf{x} \in \Omega, \\
u(\mathbf{x}, t) &= \tilde{u}_1, \ (\mathbf{x}, t) \in \partial\Omega \times [0, T],
\end{aligned}
\tag{5.6}
$$

where $\mathbf{v} \in \mathbb{R}^{2,+}, D \in \mathbb{R}^+$ and $\Omega \subset \mathbb{R}^2, T \in \mathbb{R}^+$, $\tilde{u}_0 : \Omega \to \mathbb{R}^+$ is the initial function and $\tilde{u}_1 \in \mathbb{R}^+$ is a constant. The coordinates are given as $\mathbf{x} = (x, y)^T \in \mathbb{R}$.

Application of the Lax-Wendroff scheme yields a formula accurate to the second order:

$$
\begin{aligned}
u(t^{n+1}) &\approx u(t^n) + \Delta t \left(-v_x \partial_x u(t^n) - v_y \partial_y u(t^n) \right) \\
&+ \Delta t \left(D + \frac{v_x^2 \Delta t}{2} \right) \partial_{xx} u(t^n) + \Delta t \left(D + \frac{v_y^2 \Delta t}{2} \right) \partial_{yy} u(t^n) \\
&+ \Delta t^2 v_x v_y \partial_{xy} u(t^n).
\end{aligned}
\tag{5.7}
$$

We use central differences to obtain a second-order scheme. This is similar to the 1D scheme (5.1) except for the new cross term incorporating $\partial_{xy} u$:

$$
\begin{aligned}
u_{ij}(t^{n+1}) &= u_{ij}(t^n) - \frac{\mathbf{v}_x}{2}(u_{i+1j}(t^n) - u_{i-1j}(t^n)) - \frac{\mathbf{v}_y}{2}(u_{ij+1}(t^n) - u_{ij-1}(t^n)) \\
&+ \frac{z_x \mathbf{v}_x^2}{2}(u_{i+1j}(t^n) - 2u_{ij}(t^n) + u_{i-1j}(t^n)) \\
&+ \frac{z_y \mathbf{v}_y^2}{2}(u_{ij+1}(t^n) - 2u_{ij}(t^n) + u_{ij-1}(t^n)) \\
&+ \mathbf{v}_x \mathbf{v}_y(u_{ij}(t) + u_{i-1j-1}(t^n) - u_{i-1j}(t^n) - u_{ij-1}(t^n)),
\end{aligned}
\tag{5.8}
$$

where $u_{ij} = u(x_i, y_j)$ is the solution at grid point $(x_i = i\Delta x, y_j = j\Delta y)$ with $i \in \{0, 1, \ldots, I\}, j \in \{0, 1, \ldots, J\}, I, J \in \mathbb{N}^+$ and $\Delta x, \Delta y > 0$ are the spatial step sizes in x and y directions. The time interval $[0, T]$ is given with the time points $t_n = n\Delta t$, where Δt is the equidistant time step.

We introduce the Courant-Friedrich-Levy numbers and the constants z_x, z_y given as

$$
\mathbf{v}_x = \frac{v_x \Delta t}{\Delta x}, \qquad \mathbf{v}_y = \frac{v_y \Delta t}{\Delta y},
\tag{5.9}
$$

$$
z_x = \left(\frac{2D}{\Delta t v_x^2} + 1 \right), \qquad z_y = \left(\frac{2D}{\Delta t v_y^2} + 1 \right).
\tag{5.10}
$$

This finite difference scheme is now brought into a conservation form and we have suppressed the dependencies $u_{ij}(t^n)$ on the right-hand side. The scheme is given as:

$$u_{ij}(t^{n+1}) = u_{ij} - \mathbf{v}_x(F_{i+1j} - F_{ij}) - \mathbf{v}_y(G_{ij+1} - G_{ij}), \qquad (5.11)$$

$$F_{ij} = u_{i-1j} + \frac{1}{2}\phi(\theta_{ij}^x)(1 - z_x\mathbf{v}_x)(u_{ij} - u_{i-1j}) \qquad (5.12)$$
$$\qquad - \frac{1}{2}\mathbf{v}_y(u_{i-1j} - u_{i-1j-1}),$$

$$G_{ij} = u_{ij-1} + \frac{1}{2}\phi(\theta_{ij}^y)(1 - z_y\mathbf{v}_y)(u_{ij} - u_{ij-1}) \qquad (5.13)$$
$$\qquad - \frac{1}{2}\mathbf{v}_x(u_{ij-1} - u_{i-1j-1}).$$

At this point, we introduce the flux limiter $\phi(\theta_i)$, which is used to handle steep gradients of $u(x, y, t)$ where the Lax-Wendroff scheme adds spurious oscillations (see also [57]). $\theta_i = \frac{u_{i-1} - u_{i-2}}{u_i - u_{i-1}}$ is a measure of the slope of u and is estimated in the x and y directions, respectively. For our purposes, we choose the van Leer limiter:

$$\phi(\theta) = \frac{\theta + |\theta|}{1 + |\theta|}. \qquad (5.14)$$

Remark 5.3. The limitation is important so as to have monotonous non-oscillatory schemes for high-resolution methods, see [125]. The flux and also slope limiters are discussed in [160]. In our application, we propose flux limitations.

5.1.2.3 Dimensional Splitting

In the numerical experiments, we will compare the Lax-Wendroff scheme discussed above to dimensional splitting:

$$\frac{\partial u}{\partial t} = A_x u + A_y u, \ (\mathbf{x}, t) \in \Omega \times [0, T], \qquad (5.15)$$

$$u(\mathbf{x}, t) = \tilde{u}_0(\mathbf{x}), \ \mathbf{x} \in \Omega, \qquad (5.16)$$

$$u(\mathbf{x}, t) = \tilde{u}_1, \ (\mathbf{x}, t) \in \partial\Omega \times [0, T],$$

where the operators are given by:

$$A_x = -v_x \frac{\partial}{\partial x} + D \frac{\partial^2}{\partial x^2}, \qquad (5.17)$$

$$A_y = -v_y \frac{\partial}{\partial y} + D \frac{\partial^2}{\partial y^2}, \qquad (5.18)$$

where $v_x, v_y, D \in \mathbb{R}^+$, $\tilde{u}_0 : \Omega \to \mathbb{R}^+$ is the initial function and $\tilde{u}_1 \in \mathbb{R}^+$ is a constant.

This can also be translated into a second-order finite difference scheme, filtering out the numerical viscosity $D_{num}^x = \frac{v_x\Delta x}{2}$. The notations are used as in Equation (5.8). In this case, we use an implicit BDF2 method to give a second-order dependence on time, yielding:

$$
\begin{aligned}
\frac{3}{2} u_{ij}(t^{n+1}) &= 2u_{ij}(t^n) - \frac{1}{2} u_{ij}(t^{n-1}) \\
&+ L_x[u_{ij}(t^{n+1})] + L_y[u_{ij}(t^{n+1})], \quad\quad (5.19)
\end{aligned}
$$

with spatial discretization

$$
L_x[u_{ij}(t^n)] = -\mathbf{v}_x(u_{ij}(t^n) - u_{i-1j}(t^n)) \quad\quad (5.20)
$$
$$
+ \frac{\mathbf{v}_x}{2}\left(\frac{D}{D_{num}^x} - 1\right)(u_{i+1j}(t^n) - 2u_{ij}(t^n) + u_{i-1j}(t^n)),
$$

where D_{num}^x is the numerical diffusion.

5.1.2.4 Advection-Diffusion Splitting

Another way of splitting the advection-diffusion equation is the following:

$$
\frac{\partial u}{\partial t} = Au + Bu, \ (\mathbf{x}, t) \in \Omega \times [0, T], \quad\quad (5.21)
$$
$$
u(\mathbf{x}, t) = \tilde{u}_0(\mathbf{x}), \ \mathbf{x} \in \Omega, \quad\quad (5.22)
$$
$$
u(\mathbf{x}, t) = \tilde{u}_1, \ (\mathbf{x}, t) \in \partial\Omega \times [0, T],
$$

where we assume Dirichlet boundary conditions ($\tilde{u}_1 \in \mathbb{R}^+$) and sufficiently smooth initial conditions ($\tilde{u}_0 : \Omega \to \mathbb{R}^+$), see [145].

The operators are given by:

$$
\begin{aligned}
A &= -\mathbf{v} \cdot \nabla, \\
B &= \nabla \cdot D\nabla,
\end{aligned}
$$

with constant parameters: $\mathbf{v} \in \mathbb{R}^{2,+}, D \in \mathbb{R}^+$.

The implementation of the operators is very similar to the above finite difference schemes. However, the numerical aspects demand the use of an explicit method, namely the Adams-Bashforth method or an explicit Runge-Kutta method, see [122] and [123].

Remark 5.4. Such splitting ideas can also be applied to numerical experiments, while we concentrate on each operator and its discretization method. The discretization of each operator can be done more accurately with adequate finite difference schemes, see [118]. In combination with explicit and implicit discretization schemes, see [133]. We achieve an optimal balance of the time and spatial scales. In our numerical aspects dealing with Lax-Wendroff schemes, we demand the use of explicit methods, especially the Adams-Bashforth method. For further discussions on the practicability of finite difference schemes in conjunction with iterative splitting, see [86], [91], and [102].

5.2   Domain Decomposition Methods Based on Iterative Operator Splitting Methods

In the following, we discuss an extension of the iterative splitting scheme to the spatial dimension. Such an extension generalizes the application of the iterative scheme to time and also spatial-dependent differential equations, see [91].

We concentrate on the following algorithm to iteration with fixed splitting discretization step size $\tau_n = t^{n+1} - t^n$. On the time interval $[t^n, t^{n+1}]$, we solve the following subproblems consecutively for $i = 1, 3, \ldots 2m + 1$; m is a given positive integer.

$$\frac{\partial u_i(x,t)}{\partial t} = Au_i(x,t) + Bu_{i-1}(x,t), \text{ with } u_i(t^n) = u^n, \quad (5.23)$$

$$u_1(x,t^n) = u^n, \; u_0(x,t) = 0, \text{ and}$$

$$u_i(x,t) = u_{i-1}(x,t) = \tilde{u}_1, \text{ on } \partial\Omega \times (0,T),$$

$$\frac{\partial u_{i+1}(x,t)}{\partial t} = Au_i(x,t) + Bu_{i+1}(x,t), \text{ with } u_{i+1}(x,t^n) = u^n, (5.24)$$

$$u_i(x,t) = u_{i-1}(x,t) = \tilde{u}_1, \text{ on } \partial\Omega \times (0,T),$$

where u^n is the known split approximation at time level $t = t^n$ (see [55]) and $\tilde{u}_1 \in \mathbb{R}^+$ is a given Dirichlet boundary condition.

Remark 5.5. We can generalize the iterative splitting method to a multi-iterative splitting method by introducing new splitting operators, e.g., spatial operators. Then we obtain multiple indices to control the splitting process and each iterative splitting method can be solved independently while connecting with further steps in multisplitting methods. In the following, we introduce the multi-iterative splitting method for a combined time-space splitting method.

5.2.1   Combined Time-Space Iterative Splitting Method

Notation. For the sake of simplicity and economy of space, from now on we will cease writing the dependence of the functions on the variable x. It is however important to leave the dependence on t for obvious reasons. The symbol $R_k A$ denotes the restriction of the operator A to the domain Ω_k, where k is the index of the domains, and the same notation is used for the operator B.

The following algorithm iterates with fixed splitting discretization step size τ. On the time interval $[t^n, t^{n+1}]$, we solve the following subproblems consecutively for $i = 1, 3, \ldots 2m + 1$ and $j = 1, 3, \ldots 2q + 1$, m, p are positive integers. In this notation, i represents the iteration index for the time splitting and j represents the iteration index for the spatial splitting.

$$
\begin{aligned}
\frac{\partial u_{i,j}(t)}{\partial t} &= R_1 A u_{i,j}(t) + R_2 A u_{i,j-1}(t) \\
&\quad + R_1 B u_{i-1,j}(t) + R_2 B u_{i-1,j-1}(t), \\
&\quad \text{with } u_{i,j}(t^n) = u^n,
\end{aligned}
\tag{5.25}
$$

$$
\begin{aligned}
\frac{\partial u_{i+1,j}(t)}{\partial t} &= R_1 A u_{i,j}(t) + R_2 A u_{i,j-1}(t) \\
&\quad + R_1 B u_{i+1,j}(t) + R_2 B u_{i-1,j-1}(t), \\
&\quad \text{with } u_{i+1,j}(t^n) = u^n,
\end{aligned}
\tag{5.26}
$$

$$
\begin{aligned}
\frac{\partial u_{i,j+1}(t)}{\partial t} &= R_1 A u_{i,j}(t) + R_2 A u_{i,j+1}(t) \\
&\quad + R_1 B u_{i+1,j}(t) + R_2 B u_{i-1,j-1}(t), \\
&\quad \text{with } u_{i,j+1}(t^n) = u^n,
\end{aligned}
\tag{5.27}
$$

$$
\begin{aligned}
\frac{\partial u_{i+1,j+1}(t)}{\partial t} &= R_1 A u_{i,j}(t) + R_2 A u_{i,j+1}(t) \\
&\quad + R_1 B u_{i+1,j}(t) + R_2 B u_{i+1,j+1}(t), \\
&\quad \text{with } u_{i+1,j+1}(t^n) = u^n,
\end{aligned}
\tag{5.28}
$$

where c^n is the known split approximation at time level $t = t^n$, cf. [55].

The idea of the different overlaps is presented in Figure 5.3.

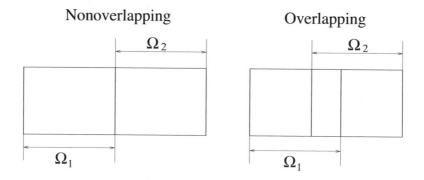

FIGURE 5.3: Graphical visualization of the overlaps.

Remark 5.6. We extend the splitting method relative to the underlying spatially discretized operators. For each subdomain, we redefine the corresponding operators in the subdomain.

5.2.2 Nonoverlapping Time-Space Iterative Splitting Method

We introduce a semidiscretization in space, where x_k are the vertices with indexing $k \in (0, \ldots, p)$ and p is the number of points. We obtain the discrete $\Omega_h = \{x_0, \ldots, x_p\}$, where we assume a uniform triangulation with grid step h. Specifically, assuming p is even, the discrete subdomains $\Omega_{1,h}$ and $\Omega_{2,h}$ consisting of the points x_k, which are associated with $k = 0, \ldots, p/2$ and $k = p/2 + 1, \ldots, p$, respectively. So, $\Omega_{1,h} \cap \Omega_{2,h} = \emptyset$ and we have the following algorithm

$$
\begin{aligned}
\frac{\partial (u_{i,j})_k(t)}{\partial t} = {} & R_{1,h} A(u_{i,j})_k(t) + R_{2,h} A(u_{i,j-1})_k(t) \\
& + R_{1,h} B(u_{i-1,j})_k(t) + R_{2,h} D(u_{i-1,j-1})_k(t), \\
& \text{with } (u_{i,j})_k(t^n) = (u^n)_k,
\end{aligned} \tag{5.29}
$$

$$
\begin{aligned}
\frac{\partial (u_{i+1,j})_k(t)}{\partial t} = {} & R_{1,h} A(u_{i,j})_k(t) + R_{2,h} A(u_{i,j-1})_k(t) \\
& + R_{1,h} B(u_{i+1,j})_k(t) + R_{2,h} B(u_{i-1,j-1})_k(t), \\
& \text{with } (u_{i+1,j})_k(t^n) = (u^n)_k,
\end{aligned} \tag{5.30}
$$

$$
\begin{aligned}
\frac{\partial (u_{i,j+1})_k(t)}{\partial t} = {} & R_{1,h} A(u_{i,j})_k(t) + R_{2,h} A(u_{i,j+1})_k(t) \\
& + R_{1,h} B(u_{i+1,j})_k(t) + R_{2,h} B(u_{i-1,j-1})_k(t), \\
& \text{with } (u_{i,j+1})_k(t^n) = (u^n)_k,
\end{aligned} \tag{5.31}
$$

$$
\begin{aligned}
\frac{\partial (u_{i+1,j+1})_k(t)}{\partial t} = {} & R_{1,h} A(u_{i,j})_k(t) + R_{2,h} A(u_{i,j+1})_k(t) \\
& + R_{1,h} B(u_{i+1,j})_k(t) + R_{2,h} B(u_{i+1,j+1})_k(t), \\
& \text{with } (u_{i+1,j+1})_k(t^n) = (u^n)_k,
\end{aligned} \tag{5.32}
$$

where u^n is the known split approximation at time level $t = t^n$.

The operators in the above equations are given as

$$
R_{1,h} A(u_{i,j})_k = \begin{cases} A u_{i,j}(x_k) & \text{for } k \in \{0, \ldots, p/2\} \\ 0 & \text{for } k \in \{p/2 + 1, \ldots, p\} \end{cases}, \tag{5.33}
$$

$$
R_{2,h} A(u_{i,j})_k = \begin{cases} 0 & \text{for } k \in \{0, \ldots, p/2\} \\ A u_{i,j}(x_k) & \text{for } k \in \{p/2, \ldots, p\} \end{cases}. \tag{5.34}
$$

The assignments for operator B are similar and given as

$$
R_{1,h} B(u_{i,j})_k = \begin{cases} B u_{i,j}(x_k) & \text{for } k \in \{0, \ldots, p/2\} \\ 0 & \text{for } k \in \{p/2 + 1, \ldots, p\} \end{cases}, \tag{5.35}
$$

$$
R_{2,h} B(u_{i,j})_k = \begin{cases} 0 & \text{for } k \in \{0, \ldots, p/2\} \\ B u_{i,j}(x_k) & \text{for } k \in \{p/2, \ldots, p\} \end{cases}. \tag{5.36}
$$

5.2.3 Overlapping Time-Space Iterative Splitting Method

We introduce a semi-discretization in space with vertices x_k, which are used as before. Now we consider the overlapping case, so we assume $\Omega_{1,h} \cap \Omega_{2,h} \neq \emptyset$. We have the following sets $\Omega_h \backslash \Omega_{2,h}$, $\Omega_{1,h} \cap \Omega_{2,h}$, and $\Omega_h \backslash \Omega_{1,h}$ consisting of the points x_k, which are associated with values of k $\{0, \ldots, p_1\}$, $\{p_1 + 1, \ldots, p_2\}$ and $\{p_2 + 1, \ldots, p\}$, respectively. We assume $p_1 < p_2 < p$ and introduce the following overlapping algorithm

$$
\begin{aligned}
\frac{\partial (u_{i,j})_k(t)}{\partial t} &= R_{1,h} A(u_{i,j})_k(t) + R_{1,2,h} A(u_{i,j}, u_{i,j-1})_k(t) \quad (5.37)\\
&+ R_{2,h} A(u_{i,j-1})_k(t) + R_{1,h} B(u_{i-1,j})_k(t)\\
&+ R_{1,2,h} B(u_{i-1,j}, u_{i-1,j-1})_k(t) + R_{2,h} B(u_{i-1,j-1})_k(t),\\
&\text{with } (u_{i,j})_k(t^n) = (u^n)_k
\end{aligned}
$$

$$
\begin{aligned}
\frac{\partial (u_{i+1,j})_k(t)}{\partial t} &= R_{1,h} A(u_{i,j})_k(t) + R_{1,2,h} A(u_{i,j}, u_{i,j-1})_k(t)\\
&+ R_{2,h} A(u_{i,j-1})_k(t) + R_{1,h} B(u_{i+1,j})_k(t)\\
&+ R_{1,2,h} B(u_{i+1,j}, u_{i-1,j-1})_k(t) + R_{2,h} B(u_{i-1,j-1})_k(t),\\
&\text{with } (u_{i+1,j})_k(t^n) = (u^n)_k \quad (5.38)
\end{aligned}
$$

$$
\begin{aligned}
\frac{\partial (u_{i,j+1})_k(t)}{\partial t} &= R_{1,h} A(u_{i,j})_k(t) + R_{1,2,h} A(u_{i,j+1}, u_{i,j})_k(t)\\
&+ R_{2,h} A(u_{i,j+1})_k(t) + R_{1,h} B(u_{i+1,j})_k(t)\\
&+ R_{1,2,h} B(u_{i+1,j}, u_{i-1,j-1})_k(t) + R_{2,h} B(u_{i-1,j-1})_k(t),\\
&\text{with } (u_{i,j+1})_k(t^n) = (u^n)_k \quad (5.39)
\end{aligned}
$$

$$
\begin{aligned}
\frac{\partial (u_{i+1,j+1})_k(t)}{\partial t} &= R_{1,h} A(u_{i,j})_k(t) + R_{1,2,h} A(u_{i,j+1}, u_{i,j})_k(t)\\
&+ R_{2,h} A(u_{i,j+1})_k(t) + R_{1,h} B(u_{i+1,j})_k(t)\\
&+ R_{1,2,h} B(u_{i+1,j}, u_{i+1,j+1})_k(t) + R_{2,h} B(u_{i+1,j+1})_k(t),\\
&\text{with } (u_{i+1,j+1})_k(t^n) = (u^n)_k \quad (5.40)
\end{aligned}
$$

where u^n is the known split approximation at time level $t = t^n$.

The assignments for operator A are given as:

$$
R_{1,h} A(u_{i,j})_k = \begin{cases} A(u_{i,j})(x_k) & \text{for } k \in \{0, \ldots, p_1\}\\ 0 & \text{for } k \in \{p_1 + 1, \ldots, p\} \end{cases}, \quad (5.41)
$$

$$
(5.42)
$$

$$
R_{1,2,h} A(u_{i,j}, u_{i,j+1})_k = \begin{cases} 0 & \text{for } k \in \{0, \ldots, p_1\}\\ A(\frac{u_{i,j}(x_k) + u_{i,j+1}(x_k)}{2}) & \text{for } k \in \{p_1 + 1, \ldots, p_2\}\\ 0 & \text{for } k \in \{p_2 + 1, \ldots, p\} \end{cases},
$$

$$
R_{2,h} A(u_{i,j})_k = \begin{cases} 0 & \text{for } k \in \{0, \ldots, p_2\}\\ A(u_{i,j}(x_k)) & \text{for } k \in \{p_2 + 1, \ldots, p\} \end{cases}. \quad (5.43)
$$

The assignments for operator B are similar and given as:

$$R_{1,h}B(u_{i,j})_k = \begin{cases} B(u_{i,j}(x_k)) & \text{for } k \in \{0,\ldots,p_1\} \\ 0 & \text{for } k \in \{p_1+1,\ldots,p\} \end{cases}, \qquad (5.44)$$

$$\qquad (5.45)$$

$$R_{1,2,h}B(u_{i,j},u_{i,j+1})_k = \begin{cases} 0 & \text{for } k \in \{0,\ldots,p_1\} \\ B(\frac{u_{i,j}(x_k)+u_{i,j+1}(x_k)}{2}) & \text{for } k \in \{p_1+1,\ldots,p_2\} \\ 0 & \text{for } k \in \{p_2+1,\ldots,p\} \end{cases},$$

$$R_{2,h}B(u_{i,j})_k = \begin{cases} 0 & \text{for } k \in \{0,\ldots,p_2\} \\ R(u_{i,j}(x_k)) & \text{for } k \in \{p_2+1,\ldots,p\} \end{cases}. \qquad (5.46)$$

The discretization of the operators is given as

$$\begin{aligned} A(u_{i,j})_k &= D/(\Delta x)^2(-(u_{i,j})_{k+1} + 2(u_{i,j})_k - (u_{i,j})_{k-1}) \\ &\quad -v/\Delta x((u_{i,j})_k - (u_{i,j})_{k-1}), \end{aligned} \qquad (5.47)$$

and

$$B(u_{i,j})_k = \lambda(u_{i,j})_k . \qquad (5.48)$$

5.2.4 Error Analysis and Convergence of Combined Method

In this section, we discuss the error analysis of our contributed combined method. We concentrate on four operators and prove the convergence with the help of generators of one-parameter C_0-semigroups (see Chapter 2 and a general introduction is given in [49]).

Theorem 5.7. *Let us consider the nonlinear operator-equation in a Banach space* **X**

$$\partial_t u(t) = A_1 u(t) + A_2 u(t) + B_1 u(t) + B_2 u(t), \quad 0 < t \le T , \quad (5.49)$$

$$u(0) = \tilde{u}_0 , \qquad (5.50)$$

where $A_1, A_2, B_1, B_2, A_1 + A_2 + B_1 + B_2 : \mathbf{X} \to \mathbf{X}$ *are given linear operators that are generators of the* C_0-semigroup *and* $\tilde{u}_0 \in \mathbf{X}$ *is a given element (initial condition). For example, we can use the operators* $A_1 = R_1 A, A_2 = R_2 A, B_1 = R_1 B, B_2 = R_2 B$, *as defined in Section 5.2.1. Then the iteration process (5.25)–(5.28) is convergent and the convergence order is one. We obtain the iterative result :* $\|e_{i,j}(t)\| \le K\tau_n \|e_{i-1,j-1}(t)\|$, *where* $\tau_n = t^{n+1} - t^n$.

Proof. Let us consider the iterations (5.25)–(5.28) on the subinterval $[t^n, t^{n+1}]$. We examine the case of exact initial conditions given as $u_{i,j}(t^n) = \tilde{u}_0$, a generalization is also possible. The error function $e_{i,j}(t) := u(t) - u_{i,j}(t)$ satisfies the relations

$$\begin{aligned}
\partial_t e_{i,j}(t) &= A_1\, e_{i,j}(t) + A_2\, e_{i,j-1}(t) + B_1\, e_{i-1,j}(t) + B_2\, e_{i-1,j-1}(t), \\
e_{i,j}(t^n) &= 0,
\end{aligned}$$
(5.51)

$$\begin{aligned}
\partial_t e_{i+1,j}(t) &= A_1\, e_{i,j}(t) + A_2\, e_{i,j-1}(t) + B_1\, e_{i+1,j}(t) + B_2\, e_{i-1,j-1}(t), \\
e_{i+1,j}(t^n) &= 0,
\end{aligned}$$
(5.52)

$$\begin{aligned}
\partial_t e_{i,j+1}(t) &= A_1\, e_{i,j}(t) + A_2\, e_{i,j+1}(t) + B_1\, e_{i+1,j}(t) + B_2\, e_{i-1,j-1}(t), \\
e_{i,j+1}(t^n) &= 0,
\end{aligned}$$
(5.53)

$$\begin{aligned}
\partial_t e_{i,j}(t) &= A_1\, e_{i,j}(t) + A_2\, e_{i,j+1}(t) + B_1\, e_{i+1,j}(t) + B_2\, e_{i+1,j+1}(t), \\
e_{i,j}(t^n) &= 0,
\end{aligned}$$
(5.54)

for $t \in [t^n, t^{n+1}]$, $i, j = 0, 2, 4, \ldots$, with $e_{1,1}(0) = 0$ and $e_{0,1}(t) = e_{1,0}(t) = e_{0,0}(t) = u(t)$.

In the following, we use the notation $\mathbf{X}^4$ for the product space $\times_{i=1}^4 \mathbf{X}$ equipped with the norm $\|(u_1, u_2, u_3, u_4)^t\| = \max_{i=1,\ldots,4}\{\|u_i\|\}$ ($u_i \in \mathbf{X}$, $i = 1, \ldots, 4$). We define the elements $\mathcal{E}_{i,j}(t)$, $\mathcal{F}_{i,j}(t) \in \mathbf{X}^4$ and the linear operator $\mathcal{A} : \mathbf{X}^4 \to \mathbf{X}^4$ by:

$$
\mathcal{E}_{i,j}(t) = \begin{bmatrix} e_{i,j}(t) \\ e_{i+1,j}(t) \\ e_{i,j+1}(t) \\ e_{i+1,j+1}(t) \end{bmatrix}, \quad
\mathcal{A} = \begin{bmatrix} A_1 & 0 & 0 & 0 \\ A_1 & B_1 & 0 & 0 \\ A_1 & B_1 & A_2 & 0 \\ A_1 & B_1 & A_2 & B_2 \end{bmatrix},
$$

$$
\mathcal{F}_{i,j}(t) = \begin{bmatrix} B_1\, e_{i-1,j}(t) + A_2\, e_{i,j-1}(t) + B_2\, e_{i-1,j-1}(t) \\ A_2\, e_{i,j-1}(t) + B_2\, e_{i-1,j-1}(t) \\ B_2\, e_{i-1,j-1}(t) \\ 0 \end{bmatrix}.
$$
(5.55)

Using notations (5.55), relations (5.51)–(5.54) can be written in the form

$$\begin{aligned}
\partial_t \mathcal{E}_{i,j}(t) &= \mathcal{A}\mathcal{E}_{i,j}(t) + \mathcal{F}_{i,j}(t), \quad t \in [t^n, t^{n+1}], \\
\mathcal{E}_{i,j}(t^n) &= 0.
\end{aligned}$$
(5.56)

Using the variations of constants formula, the solution of the abstract Cauchy problem (5.56) with a homogeneous initial condition can be written as

$$
\mathcal{E}_{i,j}(t) = \int_{t^n}^{t} \exp(\mathcal{A}(t - s))\mathcal{F}_{i,j}(s)ds, \quad t \in [t^n, t^{n+1}].
$$

(See, e.g. [49]). Hence, using the notation

$$\|\mathcal{F}_{i,j}\|_\infty = \sup_{t \in [t^n, t^{n+1}]} \|\mathcal{F}_{i,j}(t)\| \; ,$$

and taking into account Lemma 5.8 (given after this proof), which gives an estimation for $\mathcal{F}_{i,j}(t)$, we have:

$$
\begin{aligned}
\|\mathcal{E}_{i,j}(t)\| &\leq \|\mathcal{F}_{i,j}\|_\infty \int_{t^n}^t \|\exp(\mathcal{A}(t-s))\| ds \\
&\leq C \|e_{i-1,j-1}(t)\| \int_{t^n}^t \|\exp(\mathcal{A}(t-s))\| ds, \qquad (5.57)
\end{aligned}
$$

$$\text{for } t \in [t^n, t^{n+1}].$$

Owing to the linearity assumptions for the operators, $\mathcal{A}$ is a generator of the one-parameter C_0-semigroup $(\mathcal{A}(t))_{t\geq 0}$. Since $(\mathcal{A}(t))_{t\geq 0}$ is a semigroup, the so-called *growth estimation*

$$\|\exp(\mathcal{A}t)\| \leq \widetilde{K} \exp(\omega t); \quad t \geq 0 , \qquad (5.58)$$

holds for some numbers $\widetilde{K} \geq 0$ and $\omega \in \mathbb{R}$, see [49].

- Assume that $(\mathcal{A}(t))_{t\geq 0}$ is a bounded or exponentially stable semigroup, i.e., (5.58) holds for some $\omega \leq 0$. Then, obviously the estimate

$$\|\exp(\mathcal{A}t)\| \leq \widetilde{K}; \quad t \geq 0 ,$$

holds, and considering (5.57), we have

$$\|\mathcal{E}_{i,j}(t)\| \leq K\tau_n \|e_{i-1,j-1}(t)\|, \quad t \in [t^n, t^{n+1}]. \qquad (5.59)$$

- Assume that $(\mathcal{A}(t))_{t\geq 0}$ has an exponential growth for some $\omega > 0$. Integrating (5.58) yields

$$\int_{t^n}^t \|\exp(\mathcal{A}(t-s))\| ds \leq K_\omega(t), \quad t \in [t^n, t^{n+1}], \qquad (5.60)$$

where

$$K_\omega(t) = \frac{\widetilde{K}}{\omega} \left(\exp(\omega(t-t^n)) - 1\right), \quad t \in [t^n, t^{n+1}] ,$$

and hence,

$$K_\omega(t) \leq \frac{\widetilde{K}}{\omega} \left(\exp(\omega\tau_n) - 1\right) = \widetilde{K}\tau_n + \mathcal{O}(\tau_n^2) , \qquad (5.61)$$

where $\tau_n = t^{n+1} - t^n$.

The estimations (5.59), (5.60), and (5.61) result in

$$\|\mathcal{E}_{i,j}(t)\| \leq K\tau_n \|e_{i-1,j-1}(t)\|,$$

where $K = \widetilde{K} \cdot C$ in both cases.

Taking into account the definition of $\mathcal{E}_{i,j}(t)$ and the norm $\|\cdot\|$, we obtain

$$\|e_{i,j}(t)\| \leq K\tau_n \|e_{i-1,j-1}(t)\|,$$

which proves our statement.

□

Lemma 5.8. *For $\mathcal{F}_{i,j}(t)$ given by Equation (5.55), the following holds*

$$\|\mathcal{F}_{i,j}(t)\| \leq C\|e_{i-1,j-1}(t)\| .$$

Proof. We have the following norm

$$\|\mathcal{F}_{i,j}(t)\| = \max\{\|\mathcal{F}_{i,j,1}(t)\|, \|\mathcal{F}_{i,j,2}(t)\|, \|\mathcal{F}_{i,j,3}(t)\|, \|\mathcal{F}_{i,j,4}(t)\|\}.$$

Each term can be estimated as:

$$
\begin{aligned}
\|\mathcal{F}_{i,j,1}(t)\| &= \|A_2\, e_{i,j-1}(t) + B_1\, e_{i-1,j}(t) + B_2\, e_{i-1,j-1}(t)\| \leq C_1\|e_{i-1,j-1}(t)\|, \\
\|\mathcal{F}_{i,j,2}(t)\| &= \|A_2\, e_{i,j-1}(t) + B_2\, e_{i-1,j-1}(t)\| \leq C_2\|e_{i-1,j-1}(t)\|, \\
\|\mathcal{F}_{i,j,3}(t)\| &= \|B_2\, e_{i-1,j-1}(t)\| \leq C_3\|e_{i-1,j-1}(t)\|.
\end{aligned}
$$

From the theorem of Fubini, see [31], for decouplable operators, we obtain:
$\|e_{\tilde{i},\tilde{j}}(t)\| \leq \|e_{i-1,j-1}\|$, for $\tilde{i} = \{i-1, i\}$ and $\tilde{j} = \{j-1, j\}$.

Hence,

$$\|\mathcal{F}_{i,j}(t)\| \leq C\|e_{i-1,j-1}(t)\|,$$

where C is the maximum value of C_1, C_2, and C_3.

□

Remark 5.9. The results can be generalized to M decomposed domains, where M is a given positive integer. We obtain a similar result for the generalization $\mathcal{A} : \mathbf{X}^M \to \mathbf{X}^M$, see [91].

Remark 5.10. Double the number of iterations are required owing to the two partitions (i.e., i for time, j for space). Also, for higher-order accuracy, more work is needed, e.g., $2(2m+1)$ iterations for $O(\tau^{2m+1})$ accuracy.

Remark 5.11. We can generalize our results to the multidimensional case. The overlapping case is a set of the overlapping domain, which couples the multidimensional domains, see [183].

5.3 Successive Approximation for Time-Dependent Operators

The next extensions of the iterative splitting method are known as successive approximation schemes. They can be applied to time-dependent operators, we discuss the idea of the algorithm in the next paragraph, the full discussion and methods are done in the conference proceeding Geiser/Tanoglu [100].

To solve explicit time-dependent problems, the successive approximation method can be applied, which is well known, see [201].

The initial value problem is given by:

$$\frac{\partial u}{\partial t} = A(t)u(t), \ a \leq t \leq b, u(0) = u_0, \tag{5.62}$$

where $A(t)$ is a bounded time-dependent operator $A(t) \in \mathbb{R}^{M \times M}$ for all $t \in [a, b]$ and M is a given positive integer. Further, we have $a \leq b$ and $a, b \in \mathbb{R}^+$.

We rewrite:

$$\frac{\partial u}{\partial t} - A(a)u(t) + (A(t) - A(a))u(t). \tag{5.63}$$

The abstract integral is given by the so-called Duhamel Principle:

$$u(t) = \exp((t-a)A(a))u_0 + \int_a^t \exp((t-s)A(a))(A(s) - A(a))u(s) \ ds. \tag{5.64}$$

By successive approximation, we obtain:

$$u_1(t) = \exp((t-a)A(a))u_0, \tag{5.65}$$

$$\vdots$$

$$u_{n+1}(t) = \exp((t-a)A(a))u_0$$
$$+ \int_a^t \exp((t-s)A(a))(A(s) - A(a))u_n(s) \ ds, \tag{5.66}$$

and, formally, we have:

$$u(t) = \exp((t-a)A(a))u_0 + \int_a^t \exp((t-s)A(a))R(s,a)u_0 \ ds, \tag{5.67}$$

with

$$R(t,s) = \sum_{m=1}^{\infty} R_m(t,s), \tag{5.68}$$

$$R_1(t) = \begin{cases} (A(t) - A(s))\exp((t-s)A(a)) \ ds & ,s < t \\ 0 & ,s \geq t \end{cases}, \tag{5.69}$$

$$R_m(t) = \int_s^t R_1(t,\sigma)R_{m-1}(\sigma,t) \ d\sigma, \tag{5.70}$$

where we assume to compute the recursive functions R_m with quadrature rules, see also [201], and m is the number of recursions.

5.3.1   Algorithm for Successive Approximation

In this section, we construct a new numerical algorithm in order to use successive approximation as a computational tool. To illustrate how this task can be accomplished, a solution is defined for one time step, h, on the interval $[t_n, t_n + h]$

$$
u(t_n + h) \;=\; e^{hA_a} u(t_n) \tag{5.71}
$$
$$
+ \int_{t_n}^{t_n+h} e^{(t_n+h-s)A_a} (A(s) - A_a) u(s)\, ds,
$$

where $a \leq t_n \leq t \leq t_n + h \leq b$ and $A_a = A(a)$ is an $M \times M$ matrix with constant elements. Successive approximation steps can then be read as

$$
u_1(t_n + h) \;=\; e^{hA_a} u(t_n), \tag{5.72}
$$
$$
u_2(t_n + h) \;=\; e^{hA_a} u(t_n) \tag{5.73}
$$
$$
+ \int_{t_n}^{t_n+h} e^{(t_n+h-s)A_a} (A(s) - A_a) u_1(s)\, ds,
$$

$$
\vdots
$$

$$
u_k(t_n + h) \;=\; e^{hA_a} u(t_n) \tag{5.74}
$$
$$
+ \int_{t_n}^{t_n+h} e^{(t_n+h-s)A_a} (A(s) - A_a) u_{k-1}(s)\, ds.
$$

After approximating the integrals in each iteration by quadrature formulae, we re-write the solutions as

$$
u_k(t_n + h) \;=\; e^{hA_a} u(t_n) + \sum_{j=1}^{s} w_j F(c_j^*), \quad k = 2, ..., m, \tag{5.75}
$$

where $F(s) = e^{(t_n+h-s)A_a}$, w_j are weights, $c_j^* \in [t_n, t_n + h]$ are nodes of the quadrature rule and m is the number of the iterative steps.

We simply use the trapezoidal rule for approximating the integrals, we then obtain the following iterative solving scheme,

$$
u_k(t_n + h) \;=\; e^{hA_a} \left(I + \frac{h}{2} (A(t_n) - A_a) y(t_n) \right)
$$
$$
+ \frac{h}{2} (A(t_n + h) - A_a) u_{k-1}(t_n + h), \tag{5.76}
$$

for $k = 2, ...m$. Here $u(t_0) = u_0$ (initial condition), $u(t_n) = u_k(t_{n-1} + h), n =$

$1, ..N$ and $N = \frac{b-a}{h}$. The algorithm will continue until the following condition is fulfilled,

$$|u_k - u_{k-1}| \leq err, \tag{5.77}$$

where $err \in \mathbb{R}^+$ is a given error tolerance.

It can be easily seen from Equation (5.76) that the scheme involves only one approximation of the exponential of a constant matrix.

Remark 5.12. The error analysis can be done with the standard error schemes, discussed in Chapter 2. The idea is to assume a locally non-time-dependent scheme, while dealing with small time steps; see the ideas and error analysis [110].

Numerical results obtained with this algorithm are presented in the next section.

Chapter 6

Numerical Experiments

6.1 Introduction

In the numerical experiments, we discuss test examples for differential equations, a comparison to standard splitting methods, examples of extensions of the iterative splitting methods and, some real-life applications.

In the test examples, we discuss the field of application of iterative splitting schemes. First, we present scalar examples, where we discuss the important choice of how to iterate over the operators. Then, we present the field of application to vectorial examples and applications to partial differential equations. The spatial discretization is done with finite difference schemes of higher order, see [118], in such a way that we do not conflict with the order of the splitting schemes, see [82].

The results present the solutions of moderate equations with at least 2-4 iterative steps with a reduction in the size of numerical errors. Such balancing in iterative steps and time steps are discussed to reduce additional computational time.

In the comparison example, we present benchmark problems and compare them to standard Lie-Trotter, Strang splitting, and ADI (Alternating Direction Implicit) methods. We present the computational time of each scheme and its accuracy. Here are revealed the benefits of iterative schemes, which are higher-order accuracy and lower computational costs.

The extension examples show the diverse fields of application of splitting methods such as domain decomposition methods, time-dependent methods, and also nonlinear methods. Here we propose the flexibility of the iterative method, which can be modified to various applications in modern real-life problems.

We present some real-life problems, where we concentrate on two directions of the splitting schemes:

- Decoupling of a complicated real-life problem into simpler subproblems,

- Coupling of simpler subproblems with underlying software code, e.g., retardation and transport processes for complicated transport problems,

- Coupling underlying software code with special applications, e.g., reaction and transport processes for deposition and flow problems.

The first strand, which involves decoupling a problem into simpler problems, leads to the simpler equations being solved with less computational effort. Thus, embedding analytical and semianalytical solutions into simpler equations helps to accelerate the solver process.

The second strand involves coupling simpler problems and well-understood problems to more complicated problems, e.g., retardation-reaction processes with transport processes. Thus, coupling standard solver schemes with iterative splitting methods solves more complicate problems.

For both problems the main idea is a balance between a more accurate splitting method, obtained in more iterative steps after longer computational time and sufficiently accurate results with only 1-3 iterative steps. Such balancing is important and discussed. Further, larger time steps and sufficient iterative steps can be attractive to save computational time, while larger time scales are computably faster.

Iterative splitting schemes are simple to implement as a coupling method for different subproblems, even when different solver methods are used, e.g., coupling a convection-diffusion equation with a reaction equation. Here, we assume that the reaction part is solved analytically and the convection-diffusion part is solved numerically.

While the theoretical results are often restricted to Cauchy problems and coupled operators so that convergent results are obtained, practical applications of splitting schemes are more optimistic.

In practice, we can obtain higher-order results with additional iterative steps; for some examples we obtain one extra order with one extra iterative step.

Therefore, in the practical part we consider the applicability of the iterative splitting scheme as:

- Solver, for example as a flexible scheme of successive approximation methods,

- Coupling method: to couple differential equations, e.g., coupling two transport equations or continuous and discrete equations,

- Decomposition method: to separate large systems of differential equations into simpler systems or to extend it to domain decomposition methods, e.g., decomposition of large scale matrices, see [87] and [183],

- Embedded method: prestep method before a standard solver (e.g., as a predictor-corrector method), e.g., an improved initialization method to the iterative schemes, see [74] and [99],

- Generalizing standard methods, e.g., waveform relaxation methods for alternating operators.

By the way, complex models are often described with more or less understanding of the complexity of particular systems. Therefore, systematic schemes with a physically correct decomposition are important for decoupling problems into more simple, understandable models and to allow them to be coupled into more complex models in the next step. Here, the iterative splitting scheme has the advantage of leaving all operators in the equation and solving them simultaneously by implicit or explicit methods.

In this way, an understanding of partsystems is possible and the complex model is made at least partially understandable.

In this chapter, we introduce benchmark and real-life models in multi-physics problems with their physical background. We discuss the benefit of iterative splitting schemes in solving physical problems in various situations. We present different examples and discuss the results in the remarks of each subsection.

Figure 6.1 presents the numerical examples we discuss in the next sections.

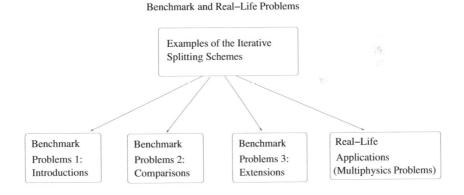

FIGURE 6.1: Numerical examples: Benchmark and real-life problems.

6.2   Benchmark Problems 1: Introduction

For the qualitative characterization of time decomposition methods, we introduce some benchmark problems.

We have chosen model problems in which the exact solutions are known, so that we can directly analyze the exact sizes of errors. Further, it is simple to compute and compare the convergence rates of the underlying schemes.

In our examples, we first consider a simple scalar equation as an ordinary differential equation (ODE) and then we consider systems of ODEs and parabolic equations. We present the flexibility and advantage of the iterative splitting method. In the schemes of various operator splitting methods, we also use the analytical method of such reduced ODEs and parabolic differential equations. We can verify the number of iteration steps relative to the order of approximation of the functions.

6.2.1   Introduction Problem 1: Starting Conditions

In the first test problem, we concentrate on the problem of the starting conditions. Here it is important which operator is dominant and taken into account as the implicit part of the iteration.

We solve an analytical benchmark problem to test the optimum iterative splitting scheme:

$$\frac{du}{dt} = \lambda u \, , \tag{6.1}$$

$$u(0) = 1 \, . \tag{6.2}$$

The split operators are given as $A = \lambda_1$, $B = \lambda_2$ where $\lambda = \lambda_1 + \lambda_2$. The exact solution is given as $u(t) = \exp(\lambda t)$.

We assume that we have extreme difference operators, e.g., $\lambda_1 << \lambda_2$. The following splitting schemes are proposed:

1. Iteration with respect to λ_1 only (scheme 1)

$$\frac{du_i}{dt} = \lambda_1 u_i + \lambda_2 u_{i-1}, \text{ for } i = 1, 2, \ldots, I, \tag{6.3}$$

$$u_i(0) = u(0),$$

where $u_0(t) = \exp(\lambda_2 t)u(0)$. Further, I is a fixed integer or a stopping criterion, e.g., $I = 5$.

2. Iteration with respect to λ_2 only (scheme 2)

$$\frac{du_i}{dt} = \lambda_1 u_{i-1} + \lambda_2 u_i, \text{ for } i = 1, 2, \ldots, I, \qquad (6.4)$$
$$u_i(0) = u(0),$$

where $u_0(t) = \exp(\lambda_2 t)u(0)$. Further, I is a fixed integer or a stopping criterion, e.g., $I = 5$.

3. Alternating iteration with respect to λ_1 and λ_2 (scheme 3)

$$\frac{du_i}{dt} = \lambda_1 u_i + \lambda_2 u_{i-1}, \qquad (6.5)$$
$$u_i(0) = u(0),$$
$$\frac{du_{i+1}}{dt} = \lambda_1 u_i + \lambda_2 u_{i+1}, \qquad (6.6)$$
$$u_{i+1}(0) = u(0),$$
$$\text{for} \qquad i = 1, 3, \ldots, I,$$

where $u_0(t) = 1$ is the starting condition.

For all iterative schemes, we assume a defined number of iterative steps, e.g., $I = 5, 6$, as a stopping criteria for the algorithm, or we assume an error condition:

$$||u_{i+1} - u_i|| \leq err, \qquad (6.7)$$

where $err \in \mathbb{R}^+$ is the error tolerance and we stop the iteration after reaching the tolerance.

In the next example, we assume to iterate only with respect to operator λ_1, meaning we apply scheme 1. Here we present a badly weighted situation.

Badly Weighted Situation for λ_1

In a situation starting with small λ_1, the iteration is badly weighted. Such problems are known as stiff problems and are also discussed in [103] and [152].

To show the blow-ups in the equations, we have the following expansion of iterated functions:

$$u_1 = 100\,e^{\frac{1}{100}\,t} - 99\,e^{\frac{1}{100}\,t}e^{-\frac{1}{100}\,t}, \tag{6.8}$$

$$u_2 = -9800\,e^{\frac{1}{100}\,t} + 99\,e^{\frac{1}{100}\,t}t + 9801\,e^{\frac{1}{100}\,t}e^{-\frac{1}{100}\,t}, \tag{6.9}$$

$$u_3 = 970300\,e^{\frac{1}{100}\,t} - 9702\,e^{\frac{1}{100}\,t}t + \frac{9801}{200}\,e^{\frac{1}{100}\,t}t^2$$
$$-970299\,e^{\frac{1}{100}\,t}e^{-\frac{1}{100}\,t}, \tag{6.10}$$

$$u_4 = -96059600\,e^{\frac{1}{100}\,t} + 960597\,e^{\frac{1}{100}\,t}t - \frac{480249}{100}\,e^{\frac{1}{100}\,t}t^2 + \frac{323433}{20000}\,e^{\frac{1}{100}\,t}t^3$$
$$+96059601\,e^{\frac{1}{100}\,t}e^{-\frac{1}{100}\,t}, \tag{6.11}$$

$$u_5 = 9509900500\,e^{\frac{1}{100}\,t} - 95099004\,e^{\frac{1}{100}\,t}t + \frac{95099103}{200}\,e^{\frac{1}{100}\,t}t^2$$
$$-\frac{15848217}{10000}\,e^{\frac{1}{100}\,t}t^3 + \frac{32019867}{8000000}\,e^{\frac{1}{100}\,t}t^4$$
$$-9509900499\,e^{\frac{1}{100}\,t}e^{-\frac{1}{100}\,t}, \tag{6.12}$$

$$u_6 = -941480149400\,e^{\frac{1}{100}\,t} + 9414801495\,e^{\frac{1}{100}\,t}t - \frac{2353700349}{50}\,e^{\frac{1}{100}\,t}t^2$$
$$+\frac{3138270399}{20000}\,e^{\frac{1}{100}\,t}t^3 - \frac{1568973483}{4000000}\,e^{\frac{1}{100}\,t}t^4 + \frac{3169966833}{4000000000}\,e^{\frac{1}{100}\,t}t^5$$
$$+941480149401\,e^{\frac{1}{100}\,t}e^{-\frac{1}{100}\,t}. \tag{6.13}$$

The Taylor expansion of the scheme (6.3) indicates large coefficients in the equation, meaning we have oscillatory and stability problems, as presented in Equations (6.8)–(6.13).

In Figure 6.2, we have computed the errors of the iteration and the exact solution. Here we obtain superlinear results with scheme 2 (6.4), that implied the iteration over the larger operator λ_2. In Figure 6.2 the underlying picks presents such superlinear convergence, which shows enormous improvement of the accuracy.

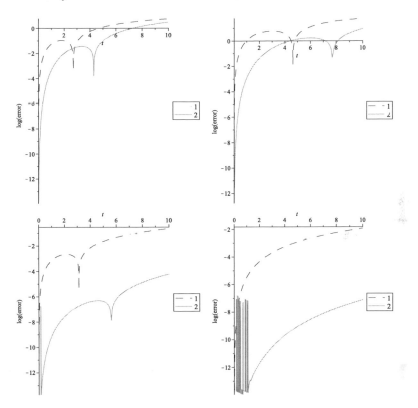

FIGURE 6.2: Experiments with different weighted parameters for the iterative schemes ([line 1]: $i = 3$, [line 2]: $i = 6$): upper left (scheme 1): $\lambda_1 = 1/100, \lambda_2 = 99/100$, upper right (scheme 3): $\lambda_1 = \lambda_2 = 1/2$, lower left (scheme 3): $\lambda_1 = 1/100, \lambda_2 = 99/100$, lower right (scheme 2): $\lambda_1 = 1/100, \lambda_2 = 99/100$, $(\log(error) = \log_{10}(\frac{|u(t)-u_i(t)|}{u(t)} \cdot 100))$.

Figure 6.3 zooms in on the analytical solution ($t \approx 5.607316185$) at the sixth iteration of scheme 3 (6.5)–(6.6) with $\lambda_1 = 1/100$ and $\lambda_2 = 99/100$. Here we have some superlinear results, while we achieve the exact solution in the iterative scheme. Such ideas can also accelerate the iterative splitting method, see [86].

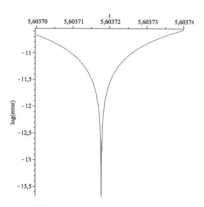

FIGURE 6.3: Zoom into the analytical solution of scheme 3 at the sixth iteration with $\lambda_1 = 1/100, \lambda_2 = 99/100$, $(\log(error) = \log_{10}(\frac{|u(t)-u_i(t)|}{u(t)} \cdot 100))$.

Remark 6.1. The numerical experiments show the important multistage situation. Starting with the badly weighted situation of a small eigenvalue, the iterative scheme shows oscillations and has weak convergence, see also [169]. Optimum balancing of the operators depends on their eigenvalues, so starting the iteration with a large eigenvalue benefits convergence, see [92].

In the following we present the next benchmark problem, which respects the stiffness of the matrices.

6.2.2   Introduction Problem 2: Stiffness of Matrices

In the following, we extend to a system of ODEs to take into account the splitting error of the scheme, while we deal with noncommuting operators; see [55].

The iterative scheme is applied to a 2×2 matrix:

$$\frac{du_1}{dt} = -\lambda_1 u_1 + \lambda_2 u_2, \tag{6.14}$$

$$\frac{du_2}{dt} = \lambda_1 u_1 - \lambda_2 u_2, \tag{6.15}$$

$$u_1(0) = 1, u_2(0) = 1. \tag{6.16}$$

The operators are given as:

$$A = \begin{pmatrix} -\lambda_1 & 0 \\ \lambda_1 & -\lambda_2 \end{pmatrix}, \tag{6.17}$$

$$B = \begin{pmatrix} 0 & \lambda_2 \\ 0 & 0 \end{pmatrix}. \tag{6.18}$$

We concentrate on the following iterative schemes and apply different iterative ideas, see also [199].

1a. Gauss-Seidel Scheme (iterations with respect to λ_1)

$$\frac{du_{1i}}{dt} = -\lambda_1 u_{1,i} + \lambda_2 u_{2,i-1}, \tag{6.19}$$

$$\frac{du_{2i}}{dt} = \lambda_1 u_{1,i} - \lambda_2 u_{2,i}, \tag{6.20}$$

$$\text{for } i = 1, 2, \ldots, I, \tag{6.21}$$

$$u_i(0) = u(0),$$

where $u_{2,0} = 0$.

1b. Gauss-Seidel Scheme (iterations with respect to λ_2)

$$\frac{du_{2i}}{dt} = -\lambda_2 u_{2,i} + \lambda_1 u_{1,i-1}, \tag{6.22}$$

$$\frac{du_{1i}}{dt} = \lambda_2 u_{2,i} - \lambda_1 u_{1,i}, \tag{6.23}$$

$$\text{for } i = 1, 2, \ldots, I, \tag{6.24}$$

$$u_i(0) = u(0),$$

where $u_{1,0} = 0$.

2. Jacobian Scheme (iterations with respect to alternating λ_1 and λ_2):

$$\frac{du_{1,i}}{dt} = -\lambda_1 u_{1,i} + \lambda_2 u_{2,i-1}, \tag{6.25}$$

$$\frac{du_{2,i}}{dt} = \lambda_1 u_{1,i} - \lambda_2 u_{2,i}, \tag{6.26}$$

$$\frac{du_{2,i+1}}{dt} = \lambda_1 u_{1,i} - \lambda_2 u_{2,i+1}, \tag{6.27}$$

$$\frac{du_{1,i+1}}{dt} = -\lambda_1 u_{1,i+1} + \lambda_2 u_{2,i+1}, \tag{6.28}$$

$$\text{for } i = 1, 3, \ldots, I, \tag{6.29}$$

$$u_i(0) = u(0),$$

where $u_{2,0} = 0$.

We assume to stop after fixed iterative steps, e.g., $I = 5, 6$, as a stopping criteria for the algorithm or we assume an error condition:

$$||u_{i+1} - u_i|| \leq err, \tag{6.30}$$

where $err \in \mathbb{R}^+$ is the error tolerance and we stop the iteration after reaching the tolerance.

We can derive the analytical solutions for the systems of such ODEs, see [47] and [48].

The solutions of the first iteration of the scheme 1a. are given as:

$$u_{11} = \exp(-\lambda_1 t), \tag{6.31}$$

$$u_{21} = \frac{\lambda_1}{\lambda_1 - \lambda_2} \exp(-\lambda_2 t) \tag{6.32}$$

$$- \frac{\lambda_1}{\lambda_1 - \lambda_2} \exp(-\lambda_1 t) + \exp(-\lambda_2 t).$$

We apply stiff cases, e.g., $\lambda_1 = 0.01$, and $\lambda_2 = 0.99$. In Figure 6.4, we have computed the errors in the Gauss-Seidel iteration scheme 1a. with the quoted parameters. With such Gauss-Seidel schemes, we achieve one convergence order more of accuracy, than with the given Jacobian scheme, see [199].

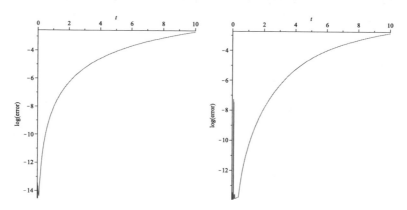

FIGURE 6.4: $\log_{10}$ error of the first (left) and second component (right) (i=4), $(\log(error) = \log_{10}(\frac{|u_j(t) - u_{j,i}(t)|}{u_j(t)} \cdot 100)$, $j = 1, 2)$.

Remark 6.2. Here, we have the benefit of starting with the optimum operators from the beginning. In our case, the scheme 1a. is more optimal, when $\lambda_1 \geq \lambda_2$, while scheme 1b. is more optimal, when $\lambda_2 \geq \lambda_1$. The errors can be reduced by solving the stiff part implicitly and the nonstiff part explicitly.

An optimal scheme is given with at least $i = 4$ and implicit iterations over the stiffer part. A compromise or mixture of both schemes are scheme 2., while respecting λ_1 in the first part and λ_2 in the second part, see [90].

6.2.3   Introduction Problem 3: Nonsplitting and Splitting

In the following, we introduce the application of splitting methods to a simple scalar equation, which can be decoupled and solved exactly as the nonsplitting version.

Here our motivation is based on the amount of computational time, which is necessary to apply the splitting scheme. Because of the iterative method, we have to investigate more computational time for this solver method.

So an improvement of iterative splitting methods is based on the following items:

- Higher accuracy with more iterative steps and larger time steps (meaning reduction of time partitions).

- Balancing of time partitions and iterative steps with respect to the computational time.

- A proposed error, e.g., $err = 10^{-3}$, helps to optimize the number of partitions and the number of iterative steps.

We must take this into account by decomposing a simple standard example. We consider the following Cauchy problem for the scalar equation:

$$\frac{du(t)}{dt} = (-\lambda_1 - \lambda_2)u(t), \ t \in [0, T], \tag{6.33}$$

$$u(0) = u_0, \tag{6.34}$$

which has the exact solution

$$u(t) = \exp(-(\lambda_1 + \lambda_2))t)u_0. \tag{6.35}$$

For problem (6.33), we split the right-hand side into the sum of two scalar operators $A + B$, where $Au = -\lambda_1 u$ and $Bu = -\lambda_2 u$. According to the iterative splitting method, (see Chapter 2), we apply the following algorithm,

$$\frac{du_i(t)}{dt} = -\lambda_1 u_i(t) - \lambda_2 u_{i-1}(t), \tag{6.36}$$

$$\frac{du_{i+1}(t)}{dt} = -\lambda_1 u_i(t) - \lambda_2 u_{i+1}(t), \tag{6.37}$$

on the interval $t \in [0, T]$, where $u_i(0) = u_{i+1}(0) = u_0$, $u_0(t) = 0$, $\forall t \in [0, T]$, and $i = 1, 3, 5, \ldots, 2m + 1$ is the number of iterations, with m a positive integer.

For the two Equations (6.36) and (6.37), we can derive the analytical solutions as

$$u_i(t) = \exp(-\lambda_1 t)\, u_i(0) + \frac{\lambda_2}{\lambda_1} u_{\text{approx,i}-1}(t)\,(\exp(-\lambda_1 t) - 1), (6.38)$$

$$u_{i+1}(t) = \frac{\lambda_1}{\lambda_2} u_{\text{approx,i}}(t)\,(\exp(-\lambda_2 t) - 1) + \exp(-\lambda_2 t)\, u_{i+1}(0), (6.39)$$

where the initial conditions are $u_{i+1}(0) = u_0$ and $u_i(0) = u_0$ with index $i = 1, 3, 5, \ldots, 2m + 1$. The time interval is $t \in [0, T]$. The starting solutions are fixed as $u_0(t) = 0$. Furthermore, $u_{\text{approx,i}-1}$ are the approximated solutions for the last iterative solution $u_{\text{i}-1}$, which has an accuracy of at least $\mathcal{O}(\tau^{2m+1})$ (with τ as the time step).

Based on these solutions, we compare the results of the iterative splitting method with the analytical solution of the complex equation, see also [82].

We perform the time discretization based on the trapezoidal rule, which is a second-order method.

The combination of handling both the iteration steps and time partitions is therefore important. We assume a time interval $[0, T]$ and divide it into n intervals of length $\tau_n = \frac{T}{n}$. We could improve the results by using smaller time steps and more iteration steps. The optimal relation is an adequately large time step with fewer iteration steps. Because of the approximation of our initial function, we can conclude that 2-4 iteration steps are sufficient, cf. Theorem 2.15.

For our example, we choose $\lambda_1 = 0.25$, $\lambda_2 = 0.5$, and $T = 1.0$, such that we obtain our exact solution with $u_{\text{exact}} = \exp(-0.75) \approx 0.4723665$.

For the simulation, we use MATLAB® 7.0 on a single CPU (2.1 GHz). In Table 6.1, we have errors at time $T = 1.0$ between the analytical and numerical results, and the computational time in $[sec]$, for the nonsplitting method. In Table 6.2 we have errors at time $T = 1.0$ and the computational time in $[sec]$, for the splitting method.

TABLE 6.1: Numerical results for the first example with a nonsplitting method and second-order trapezoidal rule.

Number of time partitions n	err $= |u_{\text{exact}} - u_{\text{num}}|$	Comput. time [sec]
100	1.5568e-12	1.82e-02
1000	1.0547e-15	1.41e-01

For the splitting method, we obtain the best results with fewer time partitions and more iteration steps, see $n = 1$ and $i = 100$, but we have to handle a long computational time. We suggest optimizing with respect to computational time, and compared to the nonsplitting method, we propose more time partitions and fewer iterations, for example $i = 2$ and $n = 10$. Comparing

TABLE 6.2: Numerical results for the first example with a splitting method and second-order trapezoidal rule.

Number of time partitions n	Number of iterations i	err $= \|u_{\text{exact}} - u_{\text{num}}\|$	Comput. time [sec]
1	2	3.8106e-02	2.83e-01
1	4	4.1633e-04	5.41e-01
1	10	5.5929e-12	1.24e+00
1	100	9.4369e-16	1.22e+01
5	2	6.1761e-03	2.80e-01
5	4	2.6127e-06	5.48e-01
5	10	1.0547e-15	1.32e+00
10	2	3.0185e-03	3.01e-01
10	4	3.1691e-07	5.85e-01
10	10	1.0547e-15	1.42e+00
100	2	2.9588e-04	6.77e-01
100	4	3.0845e-10	1.31e+00
100	10	6.6613e-16	3.21e+00

our theoretical convergence results, we obtain an order of convergence of 1 more for each iteration step, so more iteration steps result in more accurate solutions.

Remark 6.3. In this example, we show the additional work needed for the iterative method. By the way, the iterative splitting method is as accurate as the nonsplitting method. So with an optimal choice of time partitions and iterations, we can perform the splitting method and reduce computational time. For a fixed-error bound, e.g., $err = 10^{-3}$, the optimal relation is 2 iterative steps and 10 partitions or 4 iterative steps and only 1 partition. The acceleration of the scheme can also be achieved with parallel computing, in terms of more iterations. Therefore, the splitting method is as attractive as nonsplitting methods for such examples, if we restrict on a proposed and fixed-error bound.

6.2.4 Introduction Problem 4: System of ODEs with Stiff and Nonstiff Cases

Let us consider a more complicated example of these computations, where the motivation is a reversible chemical reaction between two species. We could apply this example to chemical reaction models and bio-remediation of complex processes, cf. [67]. For such problems, the time discretization schemes are important to reach higher-order results. For nonstiff problems, fast explicit Runge-Kutta methods are standard, see [122]. We concentrate on the additional computational time of the splitting scheme and its benefit of balancing time and iterative steps. Such motivations see an application field, e.g., while dealing with stiff problems and more complicate problems. Standard schemes are more or less more delicate to implement and more time consuming to solve such problems, e.g., implicit Runge-Kutta methods or IMEX (implicit-explicit) methods, see [82].

Such model problems are necessary and see the benefit of the iterative schemes.

The iterative scheme especially reduces the higher accuracy to a balance of time partitions and iterative steps.

6.2.4.1　First Experiment: Linear ODE with Nonstiff Parameters

In the first example, we deal with the following linear ordinary differential equation:

$$\frac{\partial u(t)}{\partial t} = \begin{pmatrix} -\lambda_1 & \lambda_2 \\ \lambda_1 & -\lambda_2 \end{pmatrix} u(t), \ t \in [0, T], \tag{6.40}$$

$$u(0) = u_0, \tag{6.41}$$

where the initial condition $u_0 = (1, 1)^T$ is given on the interval $[0, T]$.

The analytical solution is given by:

$$u(t) = \begin{pmatrix} c_1 - c_2 \exp\left(-(\lambda_1 + \lambda_2)t\right) \\ \frac{\lambda_1}{\lambda_2} c_1 + c_2 \exp\left(-(\lambda_1 + \lambda_2)t\right) \end{pmatrix}, \tag{6.42}$$

where

$$c_1 = \frac{2}{1 + \frac{\lambda_1}{\lambda_2}}, \quad c_2 = \frac{1 - \frac{\lambda_1}{\lambda_2}}{1 + \frac{\lambda_1}{\lambda_2}}. \tag{6.43}$$

We split our linear operator into two operators by setting

$$\frac{\partial u(t)}{\partial t} = \begin{pmatrix} -\lambda_1 & 0 \\ \lambda_1 & 0 \end{pmatrix} u(t) + \begin{pmatrix} 0 & \lambda_2 \\ 0 & -\lambda_2 \end{pmatrix} u(t). \tag{6.44}$$

We choose $\lambda_1 = 0.25$ and $\lambda_2 = 0.5$ on the interval $[0, 1]$, i.e., with $T = 1$.

We therefore have the operators:

$$A = \begin{pmatrix} -0.25 & 0 \\ 0.25 & 0 \end{pmatrix}, \quad B = \begin{pmatrix} 0 & 0.5 \\ 0 & -0.5 \end{pmatrix}. \tag{6.45}$$

For our time integration method, we assume a time interval $[0, T]$ and divide it into n intervals of length $\tau_n = \frac{T}{n}$. We can improve our results by using smaller time steps and more iteration steps.

For the initialization of our iterative method, for $i = 1$, we use $u_0(0) = (0, 0)^T$.

From the examples, we can see that the order increases with each iteration step.

In the following, we compare the results of different discretization methods for a linear ordinary differential equation. For the time discretization, we apply the second-order trapezoidal rule, the third-order backward differentiation formula (BDF3), see [122], and a Gauss-Runge-Kutta method (Gauss-RK) of at least fourth order, see [22], [23], and [122]. Thus, we obtain a maximal fourth-order method with our iterative splitting method.

Remark 6.4. So the amount of iterative steps needs higher-order time discretization schemes. But a benefit is that one can apply standard higher-order schemes, which are cheap in computational time and in implementation. For a more delicate problem, e.g., stiff operators, there is no need to concentrate on mixed time discretization schemes, e.g., IMEX methods, see [7]. Here the idea of the iterative splitting scheme is to decouple the operators with respect to their underlying timescales.

For the simulation, we use MATLAB® 7.0 on a single Linux PC with 2.1 GHz clock.

In Table 6.3, we show the errors between the analytical and numerical results, and the computational time in [*sec*], for the nonsplitting method.

Our numerical results for the splitting method are presented in Tables 6.4, 6.5, and 6.6. Additionally, we present in Table 6.4 the numerical results and computational time in [*sec*] for the nonsplitting method.

To compare the results, we choose the same number of iteration steps and time partitions for the splitting methods. The error between the analytical and numerical solutions is given as the maximum norm, i.e., $\text{err}_k = ||u_{k,\text{exact}} - u_{k,\text{num}}||$ with $k = 1, 2$.

TABLE 6.3: Numerical results for the second example with a nonsplitting method and second-order trapezoidal rule.

Number of splitting partitions n	err_1	err_2	Comput. time [sec]
100	5.1847e-13	5.1903e-13	2.27e-02
1000	2.2204e-15	3.3307e-16	1.92e-01

TABLE 6.4:   Numerical results for the second example with an
iterative splitting method and second-order trapezoidal rule.

Iteration steps i	Number of splitting partitions n	err$_1$	err$_2$	Comput. time [sec]
2	1	4.5321e-002	4.5321e-002	3.51e-01
2	10	3.9664e-003	3.9664e-003	3.93e-01
2	100	3.9204e-004	3.9204e-004	7.83e-01
3	1	7.6766e-003	7.6766e-003	5.15e-01
3	10	6.6383e-005	6.6383e-005	5.76e-01
3	100	6.5139e-007	6.5139e-007	1.15e+00
4	1	4.6126e-004	4.6126e-004	6.85e-01
4	10	4.1321e-07	4.1321e-07	7.63e-01
4	100	4.0839e-10	4.0839e-10	1.52e+00
5	1	4.6828e-005	4.6828e-005	8.51e-01
5	10	4.1382e-09	4.1382e-09	9.47e-01
5	100	4.0878e-13	4.0856e-13	1.89e+00
6	1	1.9096e-006	1.9096e-006	1.02e+00
6	10	1.7200e-11	1.7200e-11	1.13e+00
6	100	2.4425e-15	1.1102e-16	2.25e+00

A higher-order time discretization allows improved results with more iteration steps. Based on the theoretical results, we can improve the order of
the results with each iteration step. Thus, for a fourth-order time discretization, we obtain the highest order in our iterative method. Compared to the
nonsplitting method, by only applying time discretization methods, we can
reach the same accuracy but at the expense of extra computational time. An
optimal choice of fewer iterations and sufficient time partitions can save resources and we propose this as a competitive method. For large time scale
computations, which are important for such problems, we have the benefit
of achieving accurate results with sufficiently large time steps and sufficient
iterations.

The convergence results for the three methods are given in Figure 6.5.

Remark 6.5. For the nonstiff case, we obtain improved results for the iterative splitting method by increasing the number of iteration steps. Due
to improved time discretization methods, the splitting error can be reduced
with higher-order Runge-Kutta methods. Furthermore, the optimal choice
of number of iterations and sufficient time partitions can make the splitting
method more attractive than standard problems. By the way, the application
of such iterative splitting methods becomes more attractive for complicated
equations, while saving memory and computational time in the appropriate
decomposition. Larger equation systems reduce the memory and computations, while decoupling into simpler systems and balancing between time and
spatial methods.

TABLE 6.5: Numerical results for the second example with an iterative splitting method and BDF3 method.

Iteration steps i	Number of splitting partitions n	err$_1$	err$_2$
2	1	4.5321e-002	4.5321e-002
2	10	3.9664e-003	3.9664e-003
2	100	3.9204e-004	3.9204e-004
3	1	7.6766e-003	7.6766e-003
3	10	6.6385e-005	6.6385e-005
3	100	6.5312e-007	6.5312e-007
4	1	4.6126e-004	4.6126e-004
4	10	4.1334e-007	4.1334e-007
4	100	1.7864e-009	1.7863e-009
5	1	4.6833e-005	4.6833e-005
5	10	4.0122e-009	4.0122e-009
5	100	1.3737e-009	1.3737e-009
6	1	1.9040e-006	1.9040e-006
6	10	1.4350e-010	1.4336e-010
6	100	1.3742e-009	1.3741e-014

6.2.4.2 Second Experiment: Linear ODE with Stiff Parameters

In the next example, we present the benefit of the iterative splitting scheme to more delicate ODE's. Here we can decouple stiff and nonstiff parts and apply individual standard time discretization schemes.

We deal with the same equation as in the first example, now choosing $\lambda_1 = 1$ and $\lambda_2 = 10^4$ on the interval $[0,1]$.

We therefore have the operators:

$$A = \begin{pmatrix} -1 & 0 \\ 1 & 0 \end{pmatrix} \quad , \quad B = \begin{pmatrix} 0 & 10^4 \\ 0 & -10^4 \end{pmatrix}.$$

The discretization of the linear ordinary differential equation is done using the BDF3 method. Our numerical results are presented in Table 6.7. For the stiff problem, we choose more iteration steps and time partitions and show the error between the analytical and numerical solutions as the maximum norm, i.e., err$_k = \|u_{k,\text{exact}} - u_{k,\text{num}}\|$ with $k = 1, 2$.

In Table 6.7, we see that we must double the number of iteration steps to obtain the same results for the nonstiff case.

Remark 6.6. For the stiff case, we obtain improved results with more than 5 iteration steps. Because of the inexact starting function, the accuracy must

TABLE 6.6:   Numerical results for the second example with an iterative splitting method and fourth-order Gauss-RK method.

Iteration steps i	Number of splitting partitions n	err$_1$	err$_2$
2	1	4.5321e-002	4.5321e-002
2	10	3.9664e-003	3.9664e-003
2	100	3.9204e-004	3.9204e-004
3	1	7.6766e-003	7.6766e-003
3	10	6.6385e-005	6.6385e-005
3	100	6.5369e-007	6.5369e-007
4	1	4.6126e-004	4.6126e-004
4	10	4.1321e-007	4.1321e-007
4	100	4.0839e-010	4.0839e-010
5	1	4.6833e-005	4.6833e-005
5	10	4.1382e-009	4.1382e-009
5	100	4.0878e-013	4.0856e-013
6	1	1.9040e-006	1.9040e-006
6	10	1.7200e-011	1.7200e-011
6	100	2.4425e-015	1.1102e-016

be improved by more iteration steps. Higher-order time discretization methods, such as the BDF3 method and iterative splitting method, accelerate the solving process.

6.2.5   Introduction Problem 5:   Linear Partial Differential Equation

For the application to PDEs we have taken into account the spatial discretization. We apply finite difference schemes of second order or higher-order to have at least no an influence on the orders of the time and splitting schemes.

Here we have to balance spatial steps, time steps, and the iterative steps to obtain an optimal scheme with sufficient accuracy.

We consider the one-dimensional convection-diffusion-reaction equation given by

$$R\partial_t u + v\partial_x u - D\partial_{xx} u = -\lambda u \text{ , in } \Omega \times [t_0, t_{\text{end}}), \tag{6.46}$$
$$u(x, t_0) = u_{\text{exact}}(x, t_0) \text{ , in } \Omega, \tag{6.47}$$
$$u(0, t) = u_{\text{exact}}(0, t) \text{ , } u(L, t) = u_{\text{exact}}(L, t). \tag{6.48}$$

We choose $x \in [0, L] = \Omega$ with $L = 30$ and $t \in [t_0, t_{\text{end}}] = [10^4, 2 \cdot 10^4]$. Furthermore, we have constant parameters $\lambda = 10^{-5}$, $v = 0.001$, $D = 0.0001$,

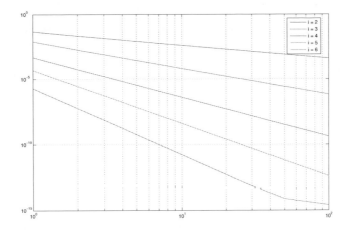

FIGURE 6.5: Convergence rates from two up to six iterations (starting with $i = 2$ at the top of the figure (lowest order) and following to $i = 6$ at the bottom of the figure (highest order)).

and $R = 1.0$. The analytical solution is given by

$$u_{\text{exact}}(x, t) = \frac{1}{2\sqrt{D\pi t}} \exp(-\frac{(x - vt)^2}{4Dt}) \exp(-\lambda t). \tag{6.49}$$

To begin away from the singular point of the exact solution, we start from the time point $t_0 = 10^4$.

Our split operators are

$$A = \frac{D}{R}\partial_{xx}, \ B = -\frac{1}{R}(\lambda + v\partial_x). \tag{6.50}$$

For the spatial discretization, we use finite differences with a spatial grid width of $\Delta x = \frac{1}{10}$ and $\Delta x = \frac{1}{100}$.

The discretization of the linear ordinary differential equation is accomplished using the BDF3 method, so we are dealing with a third-order method. Our numerical results are presented in Table 6.8. We choose different iteration steps and time partitions, and show the error between the analytical and numerical solutions as the maximum norm.

TABLE 6.7:   Numerical results for the stiff example with an iterative splitting method and BDF3 method.

Iteration steps i	Number of splitting partitions n	err_1	err_2
5	1	3.4434e-001	3.4434e-001
5	10	3.0907e-004	3.0907e-004
10	1	2.2600e-006	2.2600e-006
10	10	1.5397e-011	1.5397e-011
15	1	9.3025e-005	9.3025e-005
15	10	5.3002e-013	5.4205e-013
20	1	1.2262e-010	1.2260e-010
20	10	2.2204e-014	2.2768e-018

TABLE 6.8:   Numerical results for the third example with an iterative splitting method and BDF3 method with $\Delta x = 10^{-2}$.

Iteration steps i	Number of splitting partitions n	err at $x = 18$	err at $x = 20$	err at $x = 22$
1	10	9.8993e-002	1.6331e-001	9.9054e-002
2	10	9.5011e-003	1.6800e-002	8.0857e-003
3	10	9.6209e-004	1.9782e-002	2.2922e-004
4	10	8.7208e-004	1.7100e-002	1.5168e-005

Figure 6.6 shows the initial solution at $t_0 = 10^4$ followed by the analytical and numerical solutions at $t_{end} = 2 \cdot 10^4$, of the convection-diffusion-reaction equation.

One result that is seen is that we can reduce the error between the analytical and the numerical solutions by using more iteration steps. If we restrict ourselves to an error of 10^{-4}, we obtain an effective computation with 3 iteration steps and 10 time partitions.

Remark 6.7. For partial differential equations, we have taken into account the spatial discretization. We applied a fine grid step in the spatial discretization, such that the error of the time discretization method was dominant. We could also apply more coarser grids, if we also enlarge the time steps to balance the order of each scheme. We obtain optimal efficiency in terms of iteration steps and time partitions, if we use 10 iteration steps and 2 time partitions.

6.2.6   Introduction Problem 6: Nonlinear Ordinary Differential Equation

As a flexible method, we can also apply the iterative splitting scheme to nonlinear problems. Here we have the problem of balancing between time

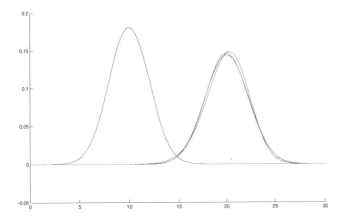

FIGURE 6.6: Initial and computed results for third example with an iterative splitting method and BDF3 method. The initial solution is plotted as the first line, the iterative solution is plotted as the second line and the exact solution is plotted as the third line.

partitions and iterative steps. Because of the linearization within the iterative scheme, a locally linear problem is solved with sufficient small time steps, see [146]. Additional iterative steps smooth the solution and obtain obtain the fixed-point solution of the iterative scheme, see [173].

As a nonlinear differential example, we choose the Bernoulli equation:

$$\frac{\partial u(t)}{\partial t} = \lambda_1 u(t) + \lambda_2 (u(t))^p, \ t \in [0, T], \tag{6.51}$$

$$u(0) = 1, \tag{6.52}$$

with the solution

$$u(t) = \left[(1 + \frac{\lambda_2}{\lambda_1}) \exp(-\lambda_1 t (1 - p)) - \frac{\lambda_2}{\lambda_1} \right]^{\frac{1}{1-p}}, \text{with } p \neq 0, 1. \tag{6.53}$$

We choose $p = 2$, $\lambda_1 = -1$, $\lambda_2 = -100$, and $\Delta x = 10^{-2}$. We divide the time interval $[0, T]$, with $T = 1$, into n intervals of length $\tau_n = \frac{T}{n}$.

We apply the iterative splitting method with the nonlinear operators

$$A(u) = \lambda_1 u(t), \ B(u) = \lambda_2 (u(t))^p. \tag{6.54}$$

The discretization of the nonlinear ordinary differential equation is performed with higher-order Runge-Kutta methods, to the third order at least, see also [72]. Our numerical results are presented in Table 6.9. We choose different iteration steps and time partitions. The error between the analytical and

TABLE 6.9: Numerical results for the
Bernoulli equation with an iterative splitting
method and BDF3 method.

Iteration steps i	Number of splitting partitions n	err
2	1	7.3724e-001
2	2	2.7910e-002
2	5	2.1306e-003
10	1	1.0578e-001
10	2	3.9777e-004
20	1	1.2081e-004
20	2	3.9782e-005

numerical solutions is shown as the maximum norm at time $T = 1.0$. The experiments have reduced errors for more iteration steps and more time partitions. Because of the time discretization method used for ODEs, we restrict the number of iteration steps to a maximum of 5 iteration steps. If we restrict the error bound to 10^{-3}, 2 iteration steps and 5 time partitions give the most effective combination.

Remark 6.8. For nonlinear ordinary differential equations, the exact starting function is an issue and important. Therefore, the initialization process is delicate and we can decrease the splitting error by using more iteration steps. Due to the linearization, we gain almost linear convergence rates. This can be improved by a higher-order linearization, see [13], [168], or a combination with Newton's method, see [147]. An overview article with an iterative splitting scheme with an embedded Newton's method is given in [105].

6.2.7 Introduction Problem 7: Coupling Convection-Diffusion and Reaction Equations with Separate Codes

The motivation in this example is code coupling. We start with two different codes, each code is solving a simpler part of a larger evolution equation. We assume to add both codes, via a splitting interface, while saving time to reprogram a full code with all equation parts. Such ideas are important to extend complicated program codes, while only single parts of the previous underlying model equation are changing, see [106].

In this example, we simulate a general species reaction transport equation, which can be written as

$$u_t + \nabla \cdot (v\,u - D\,\nabla u) = R(u) + f(x,t), \text{ in } \Omega \times (0,T), \quad (6.55)$$
$$u(x,t) = 0, \text{ on } \partial\Omega \times (0,T), \quad (6.56)$$
$$u(x,0) = u_0(x), \text{ on } \Omega, \quad (6.57)$$

where $[0, T]$ is a time interval, $x = (x_1, \ldots, x_d)^T \in \Omega$ is the space variable, and Ω is a domain in $R^d (d = 1, 2, \text{or}, 3)$. $u(x, t)$ is the unknown population density or concentration of the species, $v \in \mathbb{R}^{d,+}$ is a divergence-free velocity field, $D \in \mathbb{R}^{d,+} \times \mathbb{R}^{d,+}$ is a diffusion tensor (it is assumed that elementwise $|D_{i,j}| << |v_i| \, \forall i, j \in \{1, \ldots, d\}$), $R(u)$ is a nonlinear reaction term, and $u_0(x)$ is an initial condition.

For the reaction term, we are concerned with:

- Radioactive decay with $R(u) = -au$, as a linear reaction term

- Biodegradation model with $R(u) = au/(u + b)$, as a nonlinear reaction term.

The models and detailed discussions on biodegradation can be found in [26].

Based on the discretization methods and underlying timescales, we propose that the convection-reaction part of the equation serves as the basis for one operator and the diffusion part and right-hand side serve as the basis for a second operator. Because of the splitting, the two equations can be handled separately with the most effective discretization methods in each case. We propose for the convection-reaction part characteristic methods and a finite element method for the diffusion part.

The transport Equation (6.55) can be rewritten as an abstract operator equation:

$$u_t = A(u) + B(u), \tag{6.58}$$

with

$$A(u) = -v \cdot \nabla u + R(u), \tag{6.59}$$
$$B(u) = \nabla \cdot (D\nabla u) + f(x, t). \tag{6.60}$$

Let N be a positive integer, $\tau_n := T/N$ is the time step, and let $t_n = n\tau_n$ ($n = 0, 1, \cdots, N$) be a uniform partition of the time period $[0, T]$. We split Equation (6.55) into two equations on each small time period $[t_n, t_{n+1}]$:

$$u_t + v \cdot \nabla u = R(u), \tag{6.61}$$
$$u_t = \nabla \cdot (D\nabla u) + f(x, t). \tag{6.62}$$

For any $x \in \Omega$, the characteristic $y(s; x, t_{n+1})$ passing through (x, t_{n+1}) is determined by

$$
\begin{cases}
\dfrac{dy}{ds} = v(y, s), & s \in (t_n, t_{n+1}), \\
y(t_{n+1}; x, t_{n+1}) = x.
\end{cases} \tag{6.63}
$$

Let (x^*, t_n) be the image of exact backtracking of (x, t_{n+1}). If the characteristic reaches the boundary, then we use (x^*, t_n^*), where $x^* \in \partial\Omega$, $t_n \leq t_n^* \leq t_{n+1}$. Notice that exact tracking of characteristics is usually unavailable in practice and we have to resort to numerical means. All commonly used numerical methods for solving ODEs, e.g., Euler and Runge-Kutta methods, can be applied to problem (6.63).

Along with characteristics, the convection-reaction equation becomes a nonlinear ODE

$$\begin{cases} \dfrac{du^{(1)}}{dt} = R(u^{(1)}), \quad t \in (t_n^*, t_{n+1}), \\ u^{(1)}(x^*, t_n^*) = u^{(2)}(x^*, t_n^*), \end{cases} \tag{6.64}$$

where $u^{(2)}$ is the solution of the parabolic problem and will be explained later. When $n = 0$ or $x^* \in \partial\Omega$, we should replace $u^{(2)}(x^*, t_n^*)$ by $u_0(x^*)$ or the boundary condition.

Problem (6.64) can be solved numerically by e.g., Euler or Runge-Kutta methods. With a numerical solution, the nonlinearity in the reaction term can be well resolved if $R(u)$ is Lipschitz continuous with respect to u, which is true for biodegradation models and logistic models, see [71].

The other part of the equation to solve is an initial boundary value problem for a typical parabolic equation

$$\begin{cases} u_t^{(2)} = \nabla \cdot (D\nabla u^{(2)}) + f(x, t), \quad x \in \Omega, \ t \in (t_n, t_{n+1}), \\ u^{(2)}|_{\partial\Omega} = 0, \\ u^{(2)}(x, t_n) = u^{(1)}(x, t_{n+1}). \end{cases} \tag{6.65}$$

This problem is conventional and can be solved by finite difference, finite element, or finite volume methods.

Let Δx be the spatial mesh size in the numerical scheme (finite difference, finite element, or finite volume) for solving the parabolic part (6.65) and let τ be the temporal step size. Because $|D| << 1$, the stability condition $|D|\tau/\Delta x^2 \leq 1/2$ is readily satisfied.

6.2.7.1 First Experiment: Linear Reaction

To examine our method, we first consider a two-dimensional problem with a linear reaction term, to which we can find the exact solution. This enables us to compare the numerical and exact solutions. In particular, we have a rotating velocity $v = (-4y, 4x)^t$, a constant scalar diffusion $D > 0$, a linear reaction $R(u) = Ku$ with K a constant, and a zero source/sink, i.e., $f \equiv 0$. Assume the substance is initially normally distributed, i.e., the initial condition is specified as a Gaussian hill

$$u_0(x, y) = \exp\left(-\frac{(x - x_c)^2 + (y - y_c)^2}{2\sigma^2}\right),$$

where (x_c, y_c) is the center and $\sigma > 0$ is the standard deviation. Then the exact solution is given by

$$u(x, y, t) = \frac{2\sigma^2}{2\sigma^2 + 4Dt} \exp\left(Kt - \frac{(x^* - x_c)^2 + (y^* - y_c)^2)}{2\sigma^2 + 4Dt} \right),$$

where $(x^*, y^*, 0)$ is the backtracking foot of the characteristic from (x, y, t), that is,

$$\begin{cases} x^* = (\cos 4t)x + (\sin 4t)y, \\ y^* = -(\sin 4t)x + (\cos 4t)y. \end{cases}$$

For simplicity, we use a uniform triangular mesh. The second-order Runge-Kutta (or Heun) method is used for characteristic tracking even though exact tracking is available. The finite element solver for the parabolic part can be implemented as a dynamic link library (DLL), which is derived from the source code in an object finite element library (OFELI), see [177]. Therefore, the implementation is only a recoding and application of established Finite Element Method (FEM) libraries and can be programmed very quickly.

For numerical runs, we choose $T = \pi/2$, $\Omega = [-1, 1] \times [-1, 1]$, $D = 10^{-4}$, $K = 0.1$, $(x_c, y_c) = (-0.5, -0.5)$, and $\sigma^2 = 0.01$. For the parabolic solver, we use 20 microsteps within each global time step $[t_n, t_{n+1}]$. Accordingly, we set the maximum number of time steps in the characteristic tracking to 20. Table 6.10 lists some results for the numerical solution at the final time. We still obtain very good numerical solutions even with large global time steps.

TABLE 6.10: Numerical results of the example with a linear reaction and $\tau = \pi/8$.

Mesh size Δx	err$_{u,L_\infty}$	err$_{u,L_1}$	err$_{u,L_2}$
1/20	1.266×10^{-2}	1.247×10^{-4}	3.138×10^{-4}
1/40	1.031×10^{-2}	5.061×10^{-5}	2.085×10^{-4}
1/50	9.984×10^{-3}	4.153×10^{-5}	1.923×10^{-4}
1/60	9.796×10^{-3}	3.613×10^{-5}	1.825×10^{-4}

6.2.7.2   Second Experiment: Nonlinear Reaction

The second example is a simplified model for single species biodegradation: $R(u) = au/(u + b)$. We consider a two-dimensional problem with a constant velocity field $v = (v_1, v_2)^t$, a scalar diffusion $D > 0$, and no source/sink. The initial condition is a normal distribution (Gaussian hill). Again, we use the parabolic solver (DLL) compiled from OFELI.

For numerical runs, we choose $T = 1$, $\Omega = [-1, 1] \times [-1, 1]$, $(v_1, v_2) = (1, 1)$, $D = 10^{-4}$, $a = b = 1$, $(x_c, y_c) = (-0.5, -0.5)$, and $\sigma^2 = 0.01$. Table 6.11 lists some results of the numerical solution at the final time.

No exact solution is known for this problem. However, from Table 6.11,

TABLE 6.11: Numerical results for the
example with a nonlinear reaction.

Time step τ	Mesh size Δx	u_{min}	u_{max}
0.25	1/20	0.0	1.5159
0.25	1/40	0.0	1.5176
0.25	1/60	0.0	1.5179
0.125	1/40	0.0	1.5251
0.10	1/20	0.0	1.5248
0.10	1/50	0.0	1.5268

we observe that the operator splitting method is stable and retains a positive solution.

Remark 6.9. We apply an operator splitting method for transport equations with nonlinear reactions. In the implementation of the proposed method, we incorporate some existing commercial and free software components. By integrating the functionalities of existing software components, application developers are not obliged to create everything from scratch.

In the numerical results, we see that the method works very well and produces efficient computations. A simple extension of the proposed operator splitting method can also be made to coupled systems, see [79].

6.3 Benchmark Problems 2: Comparison with Standard Splitting Methods

In the following section, we make a comparison with standard splitting methods, such as A-B splitting, Strang splitting, and ADI methods, see [85], [133], [154], and [182]. Such benchmarks help to see the benefit of each method with respect to the underlying application. We can also position the iterative scheme in the standard methods.

When comparing the accuracy of the methods, we also compare the computational time of each method.

Such comparisons help to classify the iterative splitting methods relative to standard methods and to identify their benefits, see [98].

6.3.1 Comparison Problem 1: Iterative Splitting Method with Improved Time Discretization Methods

In the following, we compare the benefit of the iterative splitting method to standard splitting methods, such as ADI (Alternating Direction Implicit) methods. We present the benefits of higher accuracy obtained by embedding higher-order IMEX (implicit-explicit) methods and also quantify the computational burden of this improved method, see also more details in the preprint of Geiser/Tanoglu [98].

6.3.1.1 First Experiment: Heat Equation

In this section, we apply the proposed methods to the following two-dimensional problems.

We consider the following heat equation as the first test problem with initial and boundary conditions:

$$\frac{\partial u}{dt} = \frac{\partial}{\partial x} D_x \frac{\partial u}{\partial x} + \frac{\partial}{\partial y} D_y \frac{\partial u}{\partial y}, \quad \text{in } \Omega \times (0, T), \quad (6.66)$$

$$u(x, y, 0) = u_{analy}(x, y, 0), \quad \text{in } \Omega, \quad (6.67)$$

$$\frac{\partial u(x, y, t)}{\partial n} = 0, \quad \text{on } \Omega \times (0, T), \quad (6.68)$$

where $u(x, y)$ is a scalar function, $\Omega = [0, 1] \times [0, 1]$. D_x, D_y are positive constants. The analytical solution of the problem is given by

$$u_{analy}(x, y, t) = exp(-(D_x + D_y)\pi^2 t) \cos(\pi x) \cos(\pi y). \quad (6.69)$$

As an approximation error, we choose L_∞ and L_1, which are given by

$$err_{L_\infty}(\Delta t) := \max_{i=1}^{m}(\max_{j=1}^{m}(|u(x_i, y_j, t^n) - u_{analy}(x_i, y_j, t^n)|)), \quad (6.70)$$

$$err_{L_1}(\Delta t) := \sum_{i,j=1}^{m} \Delta x \Delta y |u(x_i, y_j, t^n) - u_{analy}(x_i, y_j, t^n)|, \quad (6.71)$$

and the numerical convergence rate is given as:

$$\rho_{L_1} := \frac{\log(err_{L_1(\Delta t_1)}/err_{L_1(\Delta t_2)})}{\log(\Delta t_1/\Delta t_2)} \quad (6.72)$$

same also for ρ_∞.

Our numerical results, obtained by ADI, SBDFk (Stiff Backward Differentiation Formula of k-th order), k=1,2,3, where k stands for the order of the SBDF (Stiff Backward Differentiation Formula) method, are presented in Tables 6.12, 6.13, and 6.14 for various diffusion coefficients.

First, we fixed the diffusion coefficients at $D_x = D_y = 1$ with time step $dt = 0.0005$. A comparison of L_∞, L_1 at T=0.5 and of CPU time is made in Table (6.12) for various spatial step sizes.

In the second experiment, the diffusion coefficients are fixed at $D_x = D_y = 0.001$ for the same time step. A comparison of errors L_∞, L_1 at $T = 0.5$ and of CPU time is made in Table 6.13 for various spatial steps, Δx and Δy.

In the third experiment, the diffusion coefficients are fixed at $D_x = D_y =$

TABLE 6.12:   Comparison of errors at $T = 0.5$ with various Δx and Δy when $D_x = D_y = 1$ and $dt = 0.0005$ (Standard scheme is ADI, iterative scheme is done with SBDF2-4).

	$\Delta x = \Delta y$	err_{L_∞}	err_{L_1}	ρ_{L_1}	CPU times
ADI	1/2	2.8374e-004	2.8374e-004		0.787350
	1/4	3.3299e-005	2.4260e-005	3.54	2.074755
	1/16	1.6631e-006	8.0813e-007	2.45	22.227760
SBDF2	1/2	2.811e-004	2.811e-004		0.167132
	1/4	3.2519e-005	2.3692e-005	3.568	0.339014
	1/16	1.1500e-006	5.5882e-007	2.702	2.907924
SBDF3	1/2	2.7841e-004	2.7841e-004		0.312774
	1/4	3.1721e-005	2.3111e-005	3.59	0.460088
	1/16	6.2388e-007	3.0315e-007	3.126	4.217704
SBDF4	1/2	2.7578e-004	2.7578e-004		0.400968
	1/4	3.0943e-005	2.2544e-005	3.611	0.718028
	1/16	1.1155e-007	5.4203e-008	4.348	5.550207

TABLE 6.13:   Comparison of errors at $T = 0.5$ with various Δx and Δy for $D_x = D_y = 0.001$ and $dt = 0.0005$.

	$\Delta x = \Delta y$	err_{L_∞}	err_{L_1}	CPU times
ADI	1/2	0.0019	0.0019	0.786549
	1/4	4.9226e-004	3.5864e-004	2.090480
	1/16	3.1357e-005	1.5237e-005	22.219374
SBDF2	1/2	0.0018	0.0018	0.167021
	1/4	4.8298e-004	3.5187e-004	0.341781
	1/16	2.1616e-005	1.0503e-005	2.868618
SBDF3	1/2	0.0018	0.0018	0.215563
	1/4	4.7369e-004	3.4511e-004	0.461214
	1/16	1.1874e-005	5.7699e-006	4.236695
SBDF4	1/2	0.0018	0.0018	0.274806
	1/4	4.6441e-004	3.3835e-004	0.717014
	1/16	2.1330e-006	1.0365e-006	5.517444

TABLE 6.14: Comparison of errors at $T = 0.5$ with various Δx and Δy for $D_x = D_y = 0.00001$ and $dt = 0.0005$.

	$\Delta x = \Delta y$	err_{L_∞}	err_{L_1}	CPU times
ADI	1/2	1.8694e-005	1.8694e-005	0.783630
	1/4	4.9697e-006	3.6207e-006	2.096761
	1/16	3.1665e-007	1.5386e-007	22.184733
SBDF2	1/2	1.8614e-005	1.8614e-005	0.167349
	1/4	4.8760e-006	3.5524e-006	0.342751
	1/16	2.1828e-007	1.0606e-007	2.864787
SBDF3	1/2	1.8534e-005	1.8534e-005	0.216137
	1/4	4.7823e-006	3.4842e-006	0.465666
	1/16	1.1991e-007	5.8265e-008	4.256818
SBDF4	1/2	1.8454e-005	1.8454e-005	0.399424
	1/4	4.688e-006	3.4159e-006	0.714709
	1/16	2.1539e-008	1.0466e-008	5.501323

0.00001 for the same time step. A comparison of errors L_∞, L_1 at $T = 0.5$ and of CPU time is made in Table 6.14 for various spatial steps, Δx and Δy.

In the fourth experiment, the diffusion coefficients are given as $D_x = 1$, $D_y = 0.00001$ for the same time step. A comparison of errors L_∞, L_1 at $T = 0.5$ and of CPU time is made in Table 6.15 for various spatial steps, Δx and Δy.

The Figure 6.7 presents the computations with the SBDF method, while the Figure 6.8 presents the computations with the ADI method.

Remark 6.10. In the numerical results we obtain more accurate results with the iterative schemes. By the embedded higher-order discretization schemes, we can achieve at least fourth-order iterative schemes. Also, the implementation has benefits in reducing computational time, while splitting into a cheaper explicit and more expensive implicit part.

6.3.1.2 Second Experiment: Anisotropic Equation with Time-Dependent Reaction

In a next example, we motivate with reaction processes, which are time dependent. Such processes are important to chemical reactors with different time scales for the reactive species, see [137], and [179].

It is obvious to separate into two different operators and apply dimensional splitting to the diffusion, see the motivation in [45], and [183].

The different scales of the reactive and diffusion part can be taken into account in the iterative scheme. Such splitting ideas respect the underlying time-dependent scales.

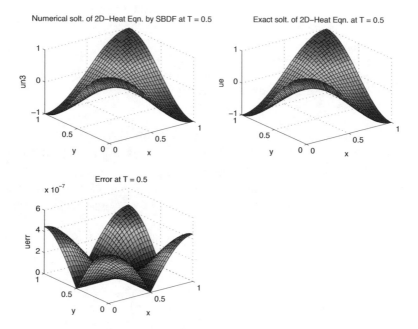

FIGURE 6.7: Comparison of errors at $T = 0.5$ with various Δx and Δy for $D_x = D_y = 0.01$ and $dt = 0.0005$.

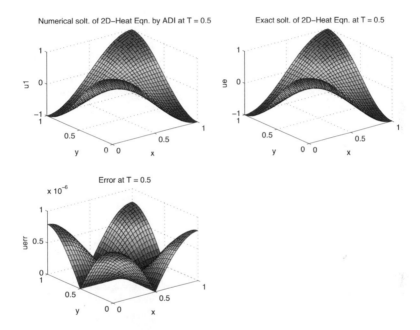

FIGURE 6.8: Comparison of errors at $T = 0.5$ with various Δx and Δy for $D_x = D_y = 0.01$ and $dt = 0.0005$.

TABLE 6.15:   Comparison of errors at $T = 0.5$ with
various Δx and Δy when $D_x = 1$, $D_y = 0.001$ and
$dt = 0.0005$.

	$\Delta x{=}\Delta y$	err_{L_∞}	err_{L_1}	CPU times
ADI	1/2	0.0111	0.0111	0.787006
	1/4	0.0020	0.0015	2.029179
	1/16	1.1426e-004	5.5520e-005	21.959890
SBDF2	1/2	0.0109	0.0109	0.210848
	1/4	0.0019	0.0014	0.385742
	1/16	2.5995e-005	1.2631e-005	2.913781
SBDF3	1/2	0.0108	0.0108	0.316777
	1/4	0.0018	0.0013	0.454392
	1/16	4.4834e-005	2.1785e-005	4.227773
SBDF4	1/2	0.0106	0.0106	0.395751
	1/4	0.0017	0.0013	0.709488
	1/16	1.1445e-004	5.5613e-005	5.562917

We deal with the following time-dependent partial differential equation:

$$\partial_t u(x,y,t) \quad = \epsilon^2 u_{xx} + u_{yy}$$
$$-(1 + \epsilon^2 + 4\epsilon^2 y^2 + 2\epsilon)e^{-t}e^{x+\epsilon y^2}, \text{ in } \Omega \times (0,T), \quad (6.73)$$
$$u(x,y,0) \quad = e^{x+\epsilon y^2}, \text{ in } \Omega, \quad (6.74)$$
$$u(x,y,t) \quad = e^{-t}e^{x+\epsilon y^2} \text{on, } \partial\Omega \times (0,T), \quad (6.75)$$

where $\Omega = [-1,1] \times [-1,1]$, $T = 0.5$, and $\epsilon \in \mathbb{R}+$ is a constant number.
We have the exact solution

$$u(x,y,t) = e^{-t}e^{x+\epsilon y^2}, \text{ in } \Omega \times (0,T). \quad (6.76)$$

The operators are split with respect to the ϵ scale and are as follows:
$$Au = \left\{ \begin{array}{l} \epsilon^2 u_{xx} - (1 + \epsilon^2 + 4\epsilon^2 y^2 + 2\epsilon)e^{-t}e^{x+\epsilon y^2} \end{array} \right. \quad \text{for } (x,y,t) \in \Omega \times (0,T),$$
and
$$Bu = \left\{ \begin{array}{l} u_{yy} \end{array} \right. \quad \text{for } (x,y,t) \in \Omega \times (0,T).$$

A comparison of the errors for $\Delta x = 2/5$ for various Δt and ϵ values with
the SBDF3 method are given in Table 6.16:

In Table 6.17, we compare the numerical solution of our second model
problem implemented according to the iterative splitting method and solved
by the trapezoidal rule, SBDF2, and SBDF3.

TABLE 6.16: Numerical results for the second problem with an iterative splitting method and SBDF3 with $\Delta x = \frac{2}{5}$.

ϵ	Δt	err_{L_∞}	err_{L_1}	CPU times
0.1	2/5	0.0542	0.0675	0.422000
	2/25	0.0039	0.0051	0.547000
	2/125	5.6742e-005	5.9181e-005	0.657000
0.001	2/5	7.6601e-004	8.6339e-004	0.610000
	2/25	4.3602e-005	5.4434e-005	0.625000
	2/125	8.0380e-006	1.0071e-005	0.703000

TABLE 6.17: Numerical results for the second problem with an iterative operator splitting method Trapezoidal rule, SBDF2, and SBDF3 with $\Delta x = \frac{2}{5}$.

	ϵ	Δt	err_{L_∞}	err_{L_1}	CPU times
Trapezoidal	1	2/25	0.9202	1.6876	0.141600
		2/125	0.8874	1.6156	0.298121
SBDF2	1	2/25	0.0473	0.0917	0.131938
		2/125	0.0504	0.0971	0.204764
SBDF3	1	2/25	0.0447	0.0878	0.149792
		2/125	0.0501	0.0964	0.537987

The exact solution, the approximate solution obtained by iterative splitting and SBDF3 and error are shown in Figure 6.9.

Remark 6.11. In the numerical example, we apply standard splitting methods and iterative splitting methods. The benefit of the iterative splitting methods is that they give more accurate solutions in less CPU time. The application of SBDF methods as standard time discretization schemes accelerates the solver process with more or less than 2 or 3 iterative steps. Thus, stiff problems can be solved by an efficient combination of the iterative splitting scheme and SBDF methods with more accuracy than the SBDF method without splitting.

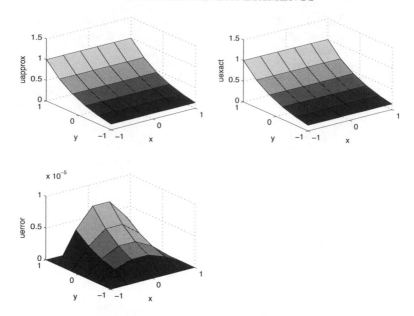

FIGURE 6.9: Exact solution, numerical solution, and error at $T = 0.5$ with various Δx and Δy for $\Delta x = \frac{2}{5}$ and $\Delta t = \frac{2}{125}$.

6.3.2   Comparison Problem 2: Iterative Splitting Method Compared to Standard Splitting Methods

In the following, we compare iterative splitting methods to standard methods, such as A-B splitting (Lie-Trotter) and Strang splitting methods.

Here, we see the benefits of iterative splitting methods as simple coding methods.

- Multistep methods can be simply applied to iterative splitting methods, while standard methods fail (because of the intermediate steps).

- Spatial discretization methods can be implemented in an iterative scheme as a locally one-dimensional method in each operator.

- In higher-order time discretization schemes, the iterative splitting method can achieve higher-order accuracy.

6.3.2.1 First Experiment: Convection-Diffusion Equation Split into its Spatial Dimensions

We deal with the two-dimensional advection-diffusion equation and periodic boundary conditions

$$\partial_t u = -v \cdot \nabla u + \nabla \cdot D \nabla u, \text{ in } \Omega \times [0, T], \tag{6.77}$$

$$= -v_x \frac{\partial u}{\partial x} - v_y \frac{\partial u}{\partial y} + D \frac{\partial^2 u}{\partial x^2} + D \frac{\partial^2 u}{\partial y^2},$$

$$u(x, t_0) = u_{analy}(x, 0), \text{ in } \Omega, \tag{6.78}$$

$$u(x, t) = u_{analy}(x, t), \text{ on } \partial\Omega \times [0, T], \tag{6.79}$$

where $\Omega = [-1, 1] \times [-1, 1]$, $T = 0.6$, $x = (x_1, x_2)^T \in \Omega$, $v = (v_x, v_y)^T \in \mathbb{R}^{2,+}$ is the velocity and we have the following constant positive parameters

$$v_x = v_y = 1,$$
$$D = 0.01,$$
$$t_0 = 0.25.$$

This advection-diffusion problem has an analytical solution

$$u_{analy}(x, t) = \frac{1}{t} \exp\left(\frac{-(x - vt)^2}{4Dt}\right),$$

which we will use as a convenient initial function:

$$u(x, t_0) = u_{analy}(x, t_0),$$

We apply dimensional splitting to our problem

$$\frac{\partial u}{\partial t} = A_x u + A_y u,$$

where

$$A_x = -v_x \frac{\partial u}{\partial x} + D \frac{\partial^2 u}{\partial x^2}.$$

We use a first-order upwind scheme for $\frac{\partial}{\partial x}$ and a second-order central difference scheme for $\frac{\partial^2}{\partial x^2}$. By introducing the artificial diffusion constant $D_x = D - \frac{v_x \Delta x}{2}$, we obtain a second-order finite difference scheme

$$L_x u(x) = -v_x \frac{u(x) - u(x - \Delta x)}{\Delta x}$$

$$+ D_x \frac{u(x + \Delta x) + u(x) + u(x - \Delta x)}{\Delta x^2}$$

because the new diffusion constant eliminates the first-order error (i.e., numerical viscosity) from the Taylor expansion of the upwind scheme. $L_y u$ is derived in the same way.

We apply a BDF5 method to gain fifth-order accuracy in time and for simplicity we suppressed the dependencies of $u(\mathbf{x}, t)$ as $u(t)$. The fifth order si given as:

$$
\begin{aligned}
L_t u(t) \quad =\; & \frac{1}{\Delta t} \left(\frac{137}{60} u(t + \Delta t) - 5u(t) + 5u(t - \Delta t) \right. \\
& \left. - \frac{10}{3} u(t - 2\Delta t) + \frac{5}{4} u(t - 3\Delta t) - \frac{1}{5} u(t - 4\Delta t) \right). \quad (6.80)
\end{aligned}
$$

Our aim is to compare the iterative splitting method to AB-splitting. Since $[A_x, A_y] = 0$, there is no splitting error for AB-splitting and therefore we cannot expect to achieve better results by iterative splitting in terms of general numerical accuracy. Instead, we will show that iterative splitting outcompetes AB-splitting in terms of computational effort and round-off error. But first, some remarks should be made concerning the special behavior of both methods when combined with high-order Runge-Kutta and BDF methods.

Splitting and schemes of higher order in time

In the following we discuss splitting schemes of higher order in time. We start with the principle of a lower order splitting scheme and later we discuss a higher-order scheme based on an iterative splitting idea.

Concerning AB-splitting: The principle of AB-splitting is well known and simple. The equation $\frac{du}{dt} = Au + Bu$ is broken down to

$$
\begin{aligned}
\frac{du^{n+1/2}}{dt} \quad &= \quad Au^{n+1/2} \\
\frac{du^{n+1}}{dt} \quad &= \quad Bu^{n+1}
\end{aligned}
$$

which are connected via $u^{n+1}(t) = u^{n+1/2}(t + \Delta t)$. This is pointed out in Figure 6.10.

AB-splitting works very well for any given one-step method like the Crank-Nicholson scheme. Not taking into account the splitting error (which is an error in time), it is also compatible with high-order schemes such as explicit-implicit Runge-Kutta schemes.

A different perspective is found if one tries to use a multistep method like the implicit BDF or the explicit Adams method with AB-splitting. These cannot be properly applied as shown by the following example:

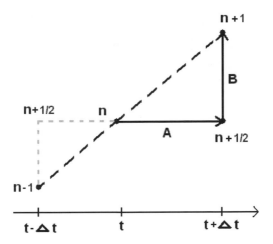

FIGURE 6.10: Principle of AB-splitting.

Choose for instance a BDF2 method, which in the case of $du/dt = f(u)$, has the scheme

$$\frac{3}{2}u(t + \Delta t) - 2u(t) + \frac{1}{2}u(t - \Delta t) = \Delta t f(u(t + \Delta t)).$$

So, the first step of AB-splitting looks like:

$$\frac{3}{2}u^{n+1/2}(t + \Delta t) - 2u^{n+1/2}(t) + \frac{1}{2}u^{n+1/2}(t - \Delta t) = \Delta t A u(t + \Delta t).$$

Here we have the problem of A-B splitting to deal with the solutions of intermediate time steps. For $u^{n+1/2}(t) = u^n(t)$ we have the solution, but we did not have the solution of $u^{n+1/2}(t - \Delta t)$. This is also shown in Figure 6.10 and it is obvious that we will have no knowledge of $u^{n+1/2}(t - \Delta t)$ unless we compute it separately, which means additional computational effort. This *overhead* increases dramatically when we move to a multistep method of higher order.

The problems mentioned with AB-splitting will not occur with a higher-order Runge-Kutta method since this only requires knowledge of $u^n(t)$.

Remarks about iterative splitting: The BDF methods apply very well to iterative splitting. Let us recall at this point that this method, although being a real splitting scheme, always remains a combination of the operators

A and B, so no steps have to be performed in one direction only. We also have a discussion, so that we will see, there is an exception to this.

In particular, we make a subdivision of our given time discretization $t_j = t_0 + j\Delta t$ into I parts. So we have subintervals $t_{j,i} = t_j + i\Delta t/I$, $0 \le i \le I$ on which we solve the following equations iteratively:

$$\frac{du^{i/I}}{dt} = Au^{i/I} + Bu^{(i-1)/I} \tag{6.81}$$

$$\frac{du^{(i+1)/I}}{dt} = Au^{i/I} + Bu^{(i+1)/I} \tag{6.82}$$

$u^{-1/I}$ is either 0 (initialization of the first iterative solution) or a reasonable approximation while $u^0 = u(t_j)$ and $u^1 = u(t_j + \Delta t)$. In fact, the order of the approximation is not of much importance if we fulfill a sufficient number of iterations. In the case $u^{-1/I} = 0$, we have the exception that a step in the A direction is taken while B is left out. The error of this step vanishes certainly after a few iterations, but mostly after just one iteration.

The crucial point here is that we only know our approximations at particular times, which happen not to be the times at which the Runge-Kutta method needs to know them. Therefore, in the case of a RK method, the values of the approximations have to be interpolated with at least the accuracy one wishes to attain with splitting and this means a lot of additional computational effort. We may summarize our results now in Table 6.18 that shows which methods are practicable for each kind of splitting scheme. In favor of the iterative splitting scheme, it should be noted that AB-splitting may be used along with the higher-order methods mentioned above but cannot maintain the order if $[A, B] \ne 0$ while iterative splitting reestablishes the maximum order of the scheme when a sufficient number of iterations are completed.

TABLE 6.18: Practicability of single-step and multistep methods (single-step methods (s.s.m.), multistep methods (m.s.m.)).

	low order s.s.m.	high order s.s.m.	m.s.m.
AB-splitting	X	X	-
Iterative splitting	X	-	X

Numerical Results

After resolving the technical aspects, we can now proceed to actual computations. A question that arises is which splitting method involves the least computational effort since we can expect them to solve the problem with more or less the same accuracy if we use practicable methods of equal order because

$[A, B] = 0$ (where 0 is the zero matrix). We tested the dimensional splitting of the 2D advection-diffusion equation with AB-splitting combined with a fifth order RK method after Dormand and Prince, see [43], and with iterative splitting in conjunction with a BDF5 scheme. We used 40×40 and 80×80 grids and completed n_t time steps with each subdivided into 10 smaller steps until we reached time $t_{end} = 0.6$, which is sufficient to see the main effects. The iterative splitting was performed in two iterations that was already sufficient to attain the desired order. In Tables 6.19 and 6.20, the errors at time t_{end} and the computation times are shown.

TABLE 6.19: Errors and computation times of AB-splitting and iterative splitting for a 40×40 grid.

Number of steps	AB error	It. spl. error	AB computation time	It. spl. computation time
5	0.1133	0.1154	0.203 s	0.141 s
10	0.1114	0.1081	0.500 s	0.312 s
30	0.1074	0.1072	1.391 s	0.907 s
50	0.1075	0.1074	2.719 s	1.594 s

TABLE 6.20: Errors and computation times of AB-splitting and iterative splitting for an 80×80 grid.

Number of steps	Error AB	Error It.spl.	AB computation time	It. spl. computation time
5	0.0288	0.0621	0.812 s	0.500 s
10	0.0276	0.0285	2.031 s	1.266 s
30	0.0268	0.0267	6.109 s	4.000 s
50	0.0265	0.0265	12.703 s	7.688 s

As we can see, the error with iterative splitting reaches the AB-splitting error after a certain number of time steps and remains below it for all additional steps. Of course, the error cannot sink below a certain value that is governed by the spatial discretization. It is to be noticed that while the computation time used for iterative splitting is always about 20-40% less than that for AB-splitting, the accuracy is, after a sufficient number of time steps, slightly better than that for AB-splitting. This is due to the round-off error, which is higher for the Runge-Kutta method because of the larger number of basic operations needed to compute the RK steps. The code for both methods is kept in the simplest form.

A future task will be to introduce noncommuting operators in order to show the superiority of iterative splitting over AB-splitting when the order in time is reduced due to the splitting error.

6.3.2.2   Second Experiment: Convection-Diffusion Equation Split into Operators

In the second experiment we consider the linear and nonlinear advection-diffusion equation in two dimensions.

Two-dimensional Advection-Diffusion Equation

We deal with the two-dimensional advection-diffusion Equation (6.77) with the parameters

$$
\begin{aligned}
v_x &= v_y = 1, \\
D &= 0.01, \\
t_0 &= 0.25.
\end{aligned}
$$

A convenient initial function is the exact solution

$$
u_{analy}(x, t) = \frac{1}{t} \exp\left(\frac{-(x - vt)^2}{4Dt}\right)
$$

at initial time t_0, $x = (x_1, x_2)^T \in \Omega$ and $v = (v_x, v_y)^T \in \mathbb{R}^{2,+}$.

We use the Lax-Wendroff scheme from Equation (5.8) with and without a flux limiter. The results are shown in Table 6.21. The values 0.0066 and 0.04

TABLE 6.21:   Errors for different diffusion constants on an 80×80 grid.

D	Ordinary LW Err.	Flux Limiter Err.
0.003	0.0230	0.0238
0.005	0.0603	0.1268
0.0066	0.0659	0.1543
0.01	unstable	0.2186
0.02	unstable	0.7331
0.04	unstable	1.3732
0.04	unstable	unstable

are rough estimates for the maximum diffusion for which the Lax-Wendroff scheme is stable, depending on the use of a flux limiter. We see that the flux limiter enhances the stability of the scheme while it leads to larger errors in regions where the ordinary scheme is stable.

Dimensional Splitting of the Two-Dimensional Advection-Diffusion

Equation

Here, the dimensional splitting from (5.19) is used in conjunction with an iterative splitting scheme. Results are shown in Table 6.22.

TABLE 6.22: Errors for different diffusion constants on an 80×80 grid.

Substeps	Iterations	$D = 0.005$	$D = 0.01$
	1	0.1760	0.0816
5	2	0.0773	0.0275
	4	0.0774	0.0275
	1	0.1733	0.0803
10	2	0.0749	0.0267
	4	0.0750	0.0267

One- and two-step iterative schemes for advection-diffusion equations

We use the splitting of (5.21) and apply one- and two-step iterative splitting.

The one-step iterative splitting algorithm is given as:

$$\frac{\partial u_i(t)}{\partial t} = Au_i + Bu_{i-1}, \tag{6.83}$$

and the two-step iterative splitting algorithm is given as:

$$\frac{\partial u_i(t)}{\partial t} = Au_i + Bu_{i-1}, \tag{6.84}$$

$$\frac{\partial u_{i+1}(t)}{\partial t} = Au_i + Bu_{i+1}, \tag{6.85}$$

respectively.

Results are shown in Tables 6.23 and 6.24. We see in the case of the one-step method that the choice of A and B makes a difference for one iteration. This is due to the dominance of the advection part. In general there is no improvement with higher iteration numbers.

In Table 6.25, we see in the case of the two-step method, that the alternation of A and B improves the result for the first iterations, but for more iterative steps, we can only improve the order with higher-order time-discretization schemes, see [72].

One- and two-step iterative schemes for the 2D Burgers equation

In our last experiment, we apply iterative splitting to the 2D Burgers equation:

TABLE 6.23: Errors for different diffusion constants on a 40×40 grid with one-step iterative splitting with $A = -\mathbf{v} \cdot \nabla$ and $B = \nabla \cdot D\nabla$.

Substeps	Iterations	$D = 0.1$	$D = 0.2$
	1	0.0362	0.0128
5	2	0.0374	0.0134
	4	0.0374	0.0134
	1	0.0361	0.0132
10	2	0.0375	0.0134
	4	0.0375	0.0134

TABLE 6.24: Errors for different diffusion constants on a 40×40 grid with one-step iterative splitting with $A = \nabla \cdot D\nabla$ and $B = -\mathbf{v} \cdot \nabla$.

Substeps	Iterations	$D = 0.1$	$D = 0.2$
	1	0.1357	0.0631
5	2	0.0375	0.0134
	4	0.0374	0.0134
	1	0.1557	0.0712
10	2	0.0376	0.0135
	4	0.0375	0.0134

TABLE 6.25: Errors for different diffusion constants on a 40×40 grid with two-step iterative splitting with $A = \nabla \cdot D\nabla$ and $B = -\mathbf{v} \cdot \nabla$.

Substeps	Iterations	$D = 0.1$	$D = 0.02$
	1	0.0246	0.0049
5	2	0.0374	0.0134
	4	0.0374	0.0134
	1	0.0128	0.0050
10	2	0.0134	0.0134
	4	0.0134	0.0134

$$\partial_t u = -\frac{1}{2}\partial_x u^2 - \frac{1}{2}\partial_y u^2 + D\partial_{xx}u + D\partial_{yy}u \text{ in } \Omega \times [0,T], \quad (6.86)$$

$$u(\mathbf{x}, 0) = u_{analy}(\mathbf{x}, 0) \text{ in } \Omega, \quad (6.87)$$

$$u(\mathbf{x}, t) = u_{analy}(\mathbf{x}, t) \text{ on } \partial\Omega \times [0,T], \quad (6.88)$$

where $u_{analy}(\mathbf{x}, t)$ is the analytical solution given in [105], $\Omega = [0, 1] \times [0, 1]$ and $T = 30$.

In this case the operators are defined as follows:

$$A = -u_{i-1}\partial_x + D\partial_{xx}, \qquad (6.89)$$
$$B = -u_{i-1}\partial_y + D\partial_{yy}, \qquad (6.90)$$

u_{i-1} denotes the $(i-1)$th iteration. This takes care of the coupling between iteration steps. We use a finite difference scheme of third order in space (central differences) and time (explicit Adams-Bashforth) and compare the results with a scheme of only second order. We also compare the iterative scheme with a standard Strang-Marchuk splitting of the second order. Numerical results are shown in Table 6.26 and 6.27.

TABLE 6.26: Burgers equation on a 40×40 grid with two-step iterative splitting and schemes of second and third order ($t_0 = 0.25$, $T = 30$).

Sub-steps	Iter-ations	Second order		Third order	
		$D = 2$	$D = 4$	$D = 2$	$D = 4$
	1	2.4188e-002	6.3023e-003	2.6636e-002	8.0688e-003
5	2	6.3253e-003	1.2224e-003	7.3006e-004	4.6906e-005
	3	6.3257e-003	1.2224e-003	7.2989e-004	4.6906e-005
	1	2.6467e-002	7.1241e-003	2.9618e-002	9.3597e-003
10	2	6.3420e-003	1.2264e-003	7.3038e-004	6.1562e-005
	3	6.3428e-003	1.2264e-003	7.2997e-004	4.6923e-005

TABLE 6.27: Burgers equation on a 40×40 grid with two-step iterative splitting and schemes of second and third order ($t_0 = 0.25$, $T = 30$).

Substeps	Iterations	Strang-Splitting	
		$D = 2$	$D = 4$
	1	4.0475e-002	3.3462e-002
5	2	-	-
	3	-	-
	1	2.7678e-002	1.7381e-002
10	2	-	-
	3	-	-

Remark 6.12. In the second experiment, we can improve iterative splitting methods by applying embedded spatial discretization methods. Higher-order schemes such as Lax-Wendroff schemes are embedded as dimensional splitting

schemes within our iterative splitting method. Nonlinear equations can also be embedded. To compute the iterative scheme, we only have to deal with spatial discretization schemes on each dimension. At least fewer iterative steps are needed to achieve higher-order results. The splitting error can be controlled and we have advantages in implementing fast time discretization schemes, while we do not consider intermediate time steps.

6.4   Benchmark Problems 3: Extensions to Iterative Splitting Methods

In this section, we present the different extensions of iterative splitting methods:

- Iterative splitting methods for time-space decomposition problems.

- Iterative splitting methods for hyperbolic equations.

- Iterative splitting methods for nonlinear equations.

- Iterative splitting methods for coupled problems.

For spatial decomposition methods, the idea is to decompose into n subdomains, where each domain is corresponding to an operator. Then we reduce exponentially the amount of computational work with operators of subdomains, see [183]. The computations can be accelerated.

For hyperbolic equations, the idea is to split into subequations and reduce the complexity of the original problem, see [92]. While each subequation can be solved faster with more adapted schemes, the amount of computation is reduced.

For nonlinear problems, we apply decomposition methods with respect to linearize the equation systems. Here splitting and linearization helps to reduce the amount of computational work, see [105].

At least for coupled problems, e.g., mobile and immobile transport regimes, we apply decomposition methods with respect to couple the equation systems. Here splitting and matrix exponential schemes help to reduce the amount of computational work.

The underlying ideas are discussed in the next examples.

6.4.1   Extension Problem 1: Spatial Decomposition Methods (Classical and Iterative Splitting Schemes)

In this section, we will present the benchmark problems for the spatial decomposition method. While classical methods like Schwarz waveform-relaxation methods embed standard splitting schemes like A-B splitting, we have considered in Section 5.2 an idea to apply an iterative splitting scheme that combines relaxation and splitting method, see also [87].

The considered subdomains are $\Omega_1 = [0, 60]$ and $\Omega_2 = [30, 150]$, as well as $\Omega_1 = [0, 100]$ and $\Omega_2 = [30, 150]$, see Figure 6.11. We apply the maximum norm, L_∞-norm, as an error norm for the error between the exact and numerical solution.

In the following section, we discuss the classical method, that includes an A-B splitting to decompose the transport equations, while the iterative

Decomposed Domains

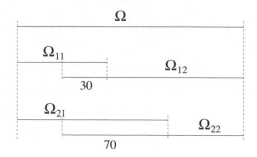

FIGURE 6.11: Overlapping situations for the numerical example.

scheme itself has a decomposition idea based on the operators A and B. So we can include a decomposition and a relaxation method into one iterative scheme, see also Section 5.2.

6.4.1.1   First Experiment: One-Dimensional Convection-Diffusion-Reaction Equation

We consider the one-dimensional convection-diffusion-reaction equation

$$\partial_t u + v\partial_x u - \partial_x D\partial_x u \;=\; -\lambda u \;,\; \text{in } \Omega \times (T_0, T_f)\,, \tag{6.91}$$

$$u(x,0) \;=\; u_{ex}(x,0)\,, \tag{6.92}$$

$$u(x,t) \;=\; u_{ex}(x,t)\,, \text{on } \partial\Omega \times (T_0, T_f)\,, \tag{6.93}$$

where $\Omega \times [T_0, T_f] = [0, 150] \times [100, 10^5]$.

The exact solution is given as

$$u_{ex}(x,t) = \frac{u_0}{2\sqrt{D\pi t}} \exp\left(-\frac{(x - vt)^2}{4Dt}\right) \exp(-\lambda t)\,. \tag{6.94}$$

The initial condition and the Dirichlet boundary conditions are defined by means of the exact solution (6.94) at starting time $T_0 = 100$ and with $u_0 = 1.0$. We have $\lambda = 10^{-5}$, $v = 0.001$, and $D = 0.0001$.

Introduction and Solution using classical methods

In order to solve the model problem using the overlapping Schwarz waveform relaxation method, we divide the domain Ω into two overlapping subdomains $\Omega_1 = [0, L_2]$ and $\Omega_2 = [L_1, L]$, where $L_1 < L_2$, and $\Omega_1 \bigcap \Omega_2 = [L_1, L_2]$ is the overlapping region for Ω_1 and Ω_2.

To start the waveform relaxation algorithm we consider first the solution of the model problem (6.91) over Ω_1 and Ω_2 as follows

$$
\begin{aligned}
v_t &= Dv_{xx} - vv_x - \lambda v \text{ over } \Omega_1 , \quad t \in [T_0, T_f], \\
v(0,t) &= f_1(t) , \quad t \in [T_0, T_f], \\
v(L_2,t) &= w(L_2,t) , \quad t \in [T_0, T_f], \\
v(x,T_0) &= u_0 \quad x \in \Omega_1,
\end{aligned}
\tag{6.95}
$$

and

$$
\begin{aligned}
w_t &= Dw_{xx} - vw_x - \lambda w \text{ over } \Omega_2 , \quad t \in [T_0, T_f], \\
w(L_1,t) &= v(L_1,t) , \quad t \in [T_0, T_f], \\
w(L,t) &= f_2(t) , \quad t \in [T_0, T_f], \\
w(x,T_0) &= u_0 \quad x \in \Omega_2,
\end{aligned}
\tag{6.96}
$$

where $v(x,t) = u(x,t)|_{\Omega_1}$ and $w(x,t) = u(x,t)|_{\Omega_2}$.

Then the Schwarz waveform relaxation is given by

$$
\begin{aligned}
v_t^{k+1} &= Dv_{xx}^{k+1} - vv_x^{k+1} - \lambda v^{k+1} \text{ over } \Omega_1 , \quad t \in [T_0, T_f], \\
v^{k+1}(0,t) &= f_1(t) , \quad t \in [T_0, T_f], \\
v^{k+1}(L_2,t) &= w^k(L_2,t) , \quad t \in [T_0, T_f], \\
v^{k+1}(x,T_0) &= u_0 \quad x \in \Omega_1,
\end{aligned}
\tag{6.97}
$$

and

$$
\begin{aligned}
w_t^{k+1} &= Dw_{xx}^{k+1} - vw_x^{k+1} - \lambda w^{k+1} \text{ over } \Omega_2 , \quad t \in [T_0, T_f], \\
w^{k+1}(L_1,t) &= v^k(L_1,t) , \quad t \in [T_0, T_f], \\
w^{k+1}(L,t) &= f_2(t) , \quad t \in [T_0, T_f], \\
w^{k+1}(x,T_0) &= u_0 \quad x \in \Omega_2.
\end{aligned}
\tag{6.98}
$$

For the solution of (6.97) and (6.98) we will apply the sequential operator splitting method (A-B splitting). For this purpose we divide each of these two equations in terms of the operators $A = D\frac{\partial^2}{\partial x^2} - v\frac{\partial}{\partial x}$ and $B = -\lambda$. The splitting scheme for each of them is given in the following form:

$$
\frac{\partial u^*(x,t)}{\partial t} = D\, u_{xx}^* - v\, u_x^* , \quad \text{with } u^*(x,t^n) = u_0 ,
\tag{6.99}
$$

$$
\frac{\partial u^{**}(x,t)}{\partial t} = -\lambda u^{**}(t) , \quad \text{with } u^{**}(x,t^n) = u^*(x,t^{n+1}) ,
\tag{6.100}
$$

where $u^*(x,t) = u^{**}(x,t) = u_1$, on $\partial\Omega \times (0,T)$, are the Dirichlet boundary conditions for the equations. The solution is given as $u(x,t^{n+1}) = u^{**}(x,t^{n+1})$. We obtain an exact method because of commuting operators.

For the discretization of equation (6.130) we apply the finite-difference method for spatial discretization and the implicit Euler method for time discretization. The discretization is given as

$$\frac{1}{t^{n+1} - t^n} \left(u^*(x_i, t^{n+1}) - u^*(x_i, t^n) \right) \tag{6.101}$$

$$= D \frac{1}{h_i^2} \left(-u^*(x_{i+1}, t^{n+1}) + 2u^*(x_i, t^{n+1}) - u^*(x_{i-1}, t^{n+1}) \right)$$

$$- v \frac{1}{h_i} \left(u^*(x_i, t^{n+1}) - u^*(x_{i-1}, t^{n+1}) \right),$$

with $u^*(x_1, t^n) = u^*(x_2, t^n) = u_0$, $u^*(x_0, t^n) = u^*(x_m, t^n) = 0$,

and $u^{**}(x, t) = \exp(-\lambda(t - t^n)\, u^*(x, t^{n+1})$, \tag{6.102}

where $h_i = x_{i+1} - x_i$ and we assume a partition with p-nodes.

We are interested in specifying the error between the solution obtained with the above-described algorithm and the exact solution. We provide a variety of results for several sizes of space- and timepartitions, and also for various overlap sizes. The time and space steps are systematically refined in order to visualize the accuracy and error reduction through the simulation over the time interval for refined time and space steps.

To be precise, we treat the cases $h = 1$, 0.5, 0.25 as spatial step-sizes, and $\Delta t = 5$, 10, 20 as time steps. The considered subdomains are $\Omega_1 = [0, 60]$, $\Omega_2 = [30, 150]$ and $\Omega_1 = [0, 100]$, $\Omega_2 = [30, 150]$, with overlap sizes 30 and 70, respectively. Both the approximated and the exact solutions are evaluated at the end time $T_f = 10^5$. The errors given in Table 6.28 are the maximum errors that occurred over the whole space domain, i.e., they are calculated by means of the maximum norm for vectors.

The approximation error is computed by the maximum error and given as $\max_{i,j} ||u_{exact}(x_i = ih, y_j = jh, T) - u_{approx}(ih, jh, T)||$.

We observe that the overlapping Schwarz waveform relaxation produces a second-order error reduction with respect to space. There is also error reduction with respect to time, but not very significant (see Table 6.28).

TABLE 6.28:   Error for the scalar convection-diffusion-reaction equation using the classical method for two different sizes of overlapping, 30 and 70.

space step	$h = 1$		$h = 0.5$		$h = 0.25$	
overlap	30	70	30	70	30	70
time step	err	err	err	err	err	err
$\Delta t = 20$	$2.81e - 3$	$2.73e - 3$	$5.22e - 4$	$5.14e - 4$	$5.66e - 4$	$4.88e - 4$
$\Delta t = 10$	$2.61e - 3$	$2.56e - 3$	$3.02e - 4$	$3.01e - 4$	$4.34e - 5$	$4.29e - 5$
$\Delta t = 5$	$2.24e - 3$	$1.28e - 3$	$2.21e - 4$	$2.20e - 4$	$1.99e - 5$	$1.97e - 5$

Solution Using the Proposed Method in Section 5.2

For the solution of (6.91)–(6.93) with the combined time-space iterative splitting method we divide the equation again in terms of the operators $A = D\frac{\partial^2}{\partial x^2} - v\frac{\partial}{\partial x}$ and $B = -\lambda$. We utilize the proposed scheme (5.38)–(5.48).

The index $k = 0, 1, \ldots p$ is associated with the subdomains, i.e., for $k = 0, \ldots, p/2$ we are working on Ω_1 and for $k = p/2+1, \ldots, p$ on Ω_2. For the first set of values for k we have actually only the effect of the restrictions of the operators A and B on Ω_1. Similarly, the second set of values for k indicates the action of the restrictions of both operators on Ω_2. The outline of the method in Section 5.2.3, which is also adopted here, is given without loss of generality for a subdomain-determining value $k = p/2$, just for an overview. This crucial value is determined appropriately according to the three cases of the overlapping subdomains, which we examine in our experiments.

The indices i and j are related to the time and space discretization, respectively. For every $k = 0, \ldots, p/2$ and for every interval of the space-discretization we solve the appropriate problems on Ω_1, for every interval of the time discretization. Similarly, for $k = p/2 + 1, \ldots, p$ on Ω_2.

From a software development point of view, the above-described numerical scheme can be realized with three "for" loops. The first, outer loop is for all values of k. After this loop there is a control for k, which distinguishes two cases for $k < p/2$ and for $k \geq p/2 + 1$, and sets up the data of the algorithm appropriately for Ω_1 or Ω_2, respectively. The second, middle loop is running for all values of i and the third, inner loop is for all values of j.

By a closer examination of the scheme (5.38)–(5.40), taking into account the definitions (5.41)–(5.46), we observe that the problems to be solved in the innermost loop are of the form $\partial_t c = Ac + Bc$, $c(x, t^n) = c^n$, where c appears with appropriate indices i and j. These problems are solved with suitable modification and implementation of the iterative splitting scheme (5.23)–(5.24). The notion of the iterative process takes place in both time- and space-dimensions.

We are interested again in specifying the error between the solution obtained with the above-described algorithm and the exact solution. We provide the same variety of results as in the previous subsection, so that a comparison between the proposed and classical methods can be established. Both the approximated and the exact solutions are evaluated at the end-time $t = 10^5$. Again, the errors given in Table 6.29 are the maximum errors that occurred over the whole space domain, i.e., they are calculated with the maximum norm for vectors.

Examining the results in Table 6.29, we notice again an error reduction for coarse space and time steps. The method attains maximum accuracy with the finest space and time refinement, and especially with the bigger size of the overlap. Comparing these results with the corresponding results presented in Table 6.28 we notice an improvement in the accuracy of the solution. It is interesting to mention that the overlapping Schwarz waveform relaxation performs slightly better than the proposed method for coarse space and time steps. After a refinement of space and time, however, the combined method is superior.

TABLE 6.29: Error for the scalar convection-diffusion-reaction equation using the proposed method for two different sizes of overlapping, 30 and 70 (with two iterative steps for time and two iterative steps for space).

space step	$h = 1$		$h = 0.5$		$h = 0.25$	
overlap	30	70	30	70	30	70
time step	err	err	err	err	err	err
$\Delta t = 20$	$4.39e - 2$	$1.20e - 2$	$5.21e - 4$	$4.53e - 4$	$5.42e - 4$	$3.21e - 4$
$\Delta t = 10$	$2.26e - 2$	$7.46e - 3$	$2.22e - 4$	$2.15e - 4$	$3.47e - 5$	$3.37e - 5$
$\Delta t = 5$	$1.47e - 2$	$3.49e - 3$	$2.13e - 4$	$1.54e - 4$	$6.49e - 6$	$8.29e - 6$

6.4.1.2 Second Experiment: Two-Dimensional Convection-Diffusion-Reaction Equation

We consider the two-dimensional convection-diffusion-reaction equation

$$\partial_t u = -v\partial_x u + \partial_x D_1 \partial_x u + \partial_y D_2 \partial_y u - \lambda u , \qquad (6.103)$$
$$\text{in } \Omega \times (T_0, T_f) ,$$

$$u(x, y, 0) = u_{ex}(x, y, 0) , \text{ in } \Omega , \qquad (6.104)$$

$$u(x, y, t) = u_{ex}(x, y, t) , \text{ on } \partial\Omega \times (T_0, T_f) , \qquad (6.105)$$

where $\Omega \times [T_0, T_f] = [0, 150] \times [0, 150] \times [100, 10^5]$.
   The exact solution is given as

$$u_{ex}(x, y, t) = \frac{u_0}{4\sqrt{D_1 \pi t}\sqrt{D_2 \pi t}} \exp\left(-\frac{(x - vt)^2}{4D_1 t}\right)$$
$$\cdot \exp(-\frac{y^2}{4D_2 t}) \exp(-\lambda t) . \qquad (6.106)$$

The initial condition and the Dirichlet boundary conditions are defined by means of the exact solution (6.106) at starting time $T_0 = 100$ and with $u_0 = 1.0$. We have $\lambda = 10^{-5}$, $v = 0.001$ and $D_1 = 0.0001$, $D_2 = 0.0005$. Both the approximated and the exact solution are evaluated at the end time $T_f = 10^5$.
   In order to develop the computer algorithms for this second example, we work in exactly the same way as in the first example. We generalize the adopted scheme for one spatial dimension of the first example to a new scheme with two spatial dimensions for the second example. The actual difference is that in this case we decompose both domains of Ω, $\Omega_x = [0, 150]$ and $\Omega_y = [0, 150]$, in two overlapping subdomains $\Omega_{x,1} = [0, L_2]$ and $\Omega_{x,2} = [L_1, L]$, where $L_1 < L_2$, and $\Omega_{x,1} \bigcap \Omega_{x,2} = [L_1, L_2]$ is the overlapping region for $\Omega_{x,1}$ and $\Omega_{x,2}$. We work similarly for $\Omega_{y,1}$ and $\Omega_{y,2}$. In order to test the algorithms, we select the same overlap sizes in both spatial dimensions x and y, which is the number that appears in the row "overlap" of the following two tables. Again, we demonstrate a comparison between the classical method combining A-B splitting with overlapping Schwarz waveform relaxation (Table 6.30) and our new proposed combined time-space iterative splitting method (Table 6.31).

The study of these tables suggests again the error reduction achieved by the two methods, and precisely the superiority of the proposed combined method over the Schwarz overlapping waveform relaxation. In contrast with the first numerical example, in this example the combined method always performs better than the classical method, even for coarse step sizes.

TABLE 6.30: Error for the second example using the classical method for two different sizes of overlapping, 30 and 70.

space step	$h = 1$		$h = 0.5$		$h = 0.25$	
overlap	30	70	30	70	30	70
time step	err	err	err	err	err	err
$\Delta t = 10$	$2.45e - 3$	$2.18e - 3$	$4.77e - 4$	$4.86e - 4$	$5.39e - 4$	$4.27e - 4$
$\Delta t = 5$	$2.32e - 3$	$2.10e - 3$	$2.76e - 4$	$2.82e - 4$	$3.98e - 5$	$3.84e - 5$
$\Delta t = 2.5$	$1.67e - 3$	$1.13e - 3$	$1.95e - 4$	$1.84e - 4$	$1.13e - 5$	$1.21e - 5$

TABLE 6.31: Error for the second example using the proposed method for two different sizes of overlapping, 30 and 70 (with two iterative steps for time and two iterative steps for space).

space step	$h = 1$		$h = 0.5$		$h = 0.25$	
overlap	30	70	30	70	30	70
time step	err	err	err	err	err	err
$\Delta t = 10$	$2.31e - 3$	$2.02e - 3$	$4.59e - 4$	$4.62e - 4$	$5.17e - 4$	$4.08e - 4$
$\Delta t = 5$	$2.19e - 3$	$1.83e - 3$	$2.54e - 4$	$2.38e - 4$	$3.74e - 5$	$3.52e - 5$
$\Delta t = 2.5$	$1.43e - 3$	$1.02e - 3$	$1.74e - 4$	$1.32e - 4$	$1.01e - 5$	$8.22e - 6$

6.4.1.3 Third Experiment: Three-Dimensional Convection-Diffusion-Reaction Equation

We consider the three-dimensional convection-diffusion-reaction equation

$$
\begin{aligned}
\partial_t u + v\partial_x u - \partial_x D_1 \partial_x u & \\
-\partial_y D_2 \partial_y u - \partial_z D_3 \partial_z u &= -\lambda u \ , \ \text{in } \Omega \times (T_0, T_f) \ , & (6.107) \\
u(x, y, z, 0) &= u_{ex}(x, y, z, 0) \ , \ \text{in } \Omega \ , & (6.108) \\
u(x, y, z, t) &= u_{ex}(x, y, z, t) \ , \text{on } \partial\Omega \times (T_0, T_f) \ , & (6.109)
\end{aligned}
$$

where $\Omega \times [T_0, T_f] = [0, 150] \times [0, 150] \times [0, 150] \times [100, 10^5]$.

The exact solution is given as

$$
\begin{aligned}
u_{ex}(x, y, z, t) &= \frac{u_0}{8\sqrt{D_1 D_2 D_3}\,(\pi t)^{3/2}} \exp\left(-\frac{(x-vt)^2}{4D_1 t}\right) \cdot \\
& \exp(-\frac{y^2}{4D_2 t}) \ \exp(-\frac{z^2}{4D_3 t}) \ \exp(-\lambda t) \ . & (6.110)
\end{aligned}
$$

The initial condition and the Dirichlet boundary conditions are defined with the exact solution (6.110) at starting time $T_0 = 100$ and with $u_0 = 1.0$. We have $\lambda = 10^{-5}$, $v = 0.001$ and $D_1 = 0.0001$, $D_2 = 0.0005$, $D_3 = 0.0008$. Both the approximated and the exact solutions are evaluated at the end time $T_f = 10^5$.

The design of the computer algorithms for this third example is done in just the same way as in the first two examples. We generalize the adopted scheme for one spatial dimension of the first example to a new scheme with three spatial dimensions for the third example. The actual difference is that in this case we decompose all three domains of Ω, $\Omega_x = [0, 150]$, $\Omega_y = [0, 150]$ and $\Omega_z = [0, 150]$, in two overlapping subdomains $\Omega_{x,1} = [0, L_2]$ and $\Omega_{x,2} = [L_1, L]$, where $L_1 < L_2$, and $\Omega_{x,1} \cap \Omega_{x,2} = [L_1, L_2]$ is the overlapping region for $\Omega_{x,1}$ and $\Omega_{x,2}$. We work similarly for $\Omega_{y,1}$ and $\Omega_{y,2}$, as for $\Omega_{z,1}$ and $\Omega_{z,2}$. In order to test the algorithms, we select for the sake of simplicity the same overlap sizes in all the spatial dimensions x, y, and z, which is the number that appears in the row "overlap" of the following two tables. Again, we demonstrate a comparison between the classical method combining A-B splitting with overlapping Schwarz waveform relaxation (Table 6.32) and our new proposed combined time-space iterative splitting method (Table 6.33). These tables reveal the same convergence and accuracy properties as Tables 6.30 and 6.31.

TABLE 6.32:   Error for the third example using the classical method for two different sizes of overlapping 30 and 70.

space step	$h = 1$		$h = 0.5$		$h = 0.25$	
overlap	30	70	30	70	30	70
time step	err	err	err	err	err	err
$\Delta t = 20$	$3.51e-3$	$3.47e-3$	$3.45e-3$	$3.42e-4$	$3.21e-3$	$2.99e-4$
$\Delta t = 10$	$3.40e-3$	$3.31e-3$	$3.24e-3$	$3.21e-3$	$3.19e-3$	$2.93e-4$
$\Delta t = 5$	$3.39e-3$	$3.26e-3$	$3.22e-3$	$3.16e-3$	$3.08e-3$	$2.90e-4$

TABLE 6.33:   Error for the third example using the proposed method for two different sizes of overlapping, 30 and 70 (with two iterative steps for time and two iterative steps for space).

space step	$h = 1$		$h = 0.5$		$h = 0.25$	
overlap	30	70	30	70	30	70
time step	err	err	err	err	err	err
$\Delta t = 20$	$3.50e-3$	$3.45e-3$	$3.22e-3$	$3.10e-3$	$2.20e-3$	$2.03e-3$
$\Delta t = 10$	$3.25e-3$	$3.31e-3$	$3.01e-3$	$2.92e-4$	$2.79e-4$	$2.53e-4$
$\Delta t = 5$	$3.01e-3$	$2.76e-4$	$2.32e-4$	$2.06e-4$	$1.88e-5$	$1.52e-5$

6.4.1.4 Fourth Experiment: Time-Dependent Diffusion Equation

In the fourth example we deal with the following time-dependent partial differential equation in two dimensions (see also [4]),

$$
\begin{aligned}
\partial_t u(x, y, t) &= u_{xx} + u_{yy} - 4(1 + y^2)e^{-t}e^{x+y^2}, \text{ in } \Omega \times [0, T], &(6.111)\\
u(x, y, 0) &= e^{x+y^2}, \text{ in } \Omega, &(6.112)\\
u(x, y, t) &= e^{-t}e^{x+y^2}, \text{ on } \partial\Omega \times [0, T], &(6.113)
\end{aligned}
$$

with exact solution

$$
u(x, y, t) = e^{-t}e^{x+y^2}, \text{ in } \Omega \times [0, T]. \tag{6.114}
$$

The domain is given as $\Omega = [-1, 1] \times [-1, 1]$ and the time interval is $[0, T] = [0, 1]$ and again use higher-order finite difference schemes in time and space. The operators used for the splitting methods are

$$
Au = \begin{cases} u_{xx} + u_{yy} - 4(1 + y^2)e^{-t}e^{x+y^2}, & \text{for } (x, y) \in \Omega_1 \\ 0, & \text{for } (x, y) \in \Omega_2 \end{cases},
$$

and

$$
Bu = \begin{cases} 0, & \text{for } (x, y) \in \Omega_1 \\ u_{xx} + u_{yy} - 4(1 + y^2)e^{-t}e^{x+y^2}, & \text{for } (x, y) \in \Omega_2 \end{cases},
$$

with $\Omega_1 = [-1, 1] \times [-1, 0]$ and $\Omega_2 = [-1, 1] \times [0, 1]$.

The approximation error is computed by the maximum error and given as $\max_{i,j} \|u_{exact}(x_i = ih, y_j = jh, T) - u_{approx}(ih, jh, T)\|$.

Again, we discuss a comparison between the classical method combining A-B splitting with overlapping Schwarz waveform relaxation (Table 6.34) and our new proposed combined time-space iterative splitting method (Table 6.35). These tables reveal the same convergence and accuracy properties as Tables 6.32 and 6.33.

TABLE 6.34: Error for the fourth example using the classical method for two different sizes of overlapping, 30 and 70.

space step	$h = 1$		$h = 0.5$		$h = 0.25$	
overlap	30	70	30	70	30	70
time step	err	err	err	err	err	err
$\Delta t = 20$	$2.56e - 3$	$4.14e - 4$	$2.19e - 4$	$9.91e - 5$	$9.67e - 5$	$9.46e - 6$
$\Delta t = 10$	$2.44e - 3$	$7.64e - 4$	$5.28e - 4$	$9.72e - 5$	$9.50e - 5$	$9.30e - 6$
$\Delta t = 5$	$1.54e - 3$	$1.45e - 4$	$9.64e - 5$	$1.32e - 5$	$9.88e - 6$	$3.57e - 6$

We observe an error reduction for each time and space refinement, which leads finally to a very satisfying accuracy. Furthermore, in comparison with previous tables, we see in Table 6.35 that the overlap plays a more important role in the error reduction.

TABLE 6.35: Error for the fourth example using the proposed method for two different sizes of overlapping, 30 and 70 (with two iterative steps for time and two iterative steps for space).

space step	$h = 1$		$h = 0.5$		$h = 0.25$	
overlap	30	70	30	70	30	70
time step	err	err	err	err	err	err
$\Delta t = 20$	$1.41e-3$	$9.12e-4$	$8.91e-4$	$9.72e-5$	$9.56e-5$	$9.23e-6$
$\Delta t = 10$	$1.32e-3$	$5.06e-4$	$4.85e-4$	$9.61e-5$	$9.44e-5$	$9.19e-6$
$\Delta t = 5$	$1.12e-3$	$1.18e-4$	$9.21e-5$	$8.95e-6$	$8.59e-6$	$8.97e-7$

Remark 6.13. We observed in the four experiments an error reduction for time and spatial refinement. The proposed iterative method can improve the accuracy of the results by increasing the overlapping area. The effectiveness of our method can be achieved by optimizing the number of iterative steps and the number of time steps. The control of the overlap leads to a robust and attractive method, while reducing the error with each iteration step (see [66]).

Remark 6.14. The efficient results are taken with the spatial and time decomposition methods. While in the standard method, the spatial scale is splitted by a Schwarz waveform relaxation scheme and additional timescale with an A-B or Strang splitting method, the iterative splitting scheme includes both time and spatial decomposition methods. The numerical methods show the extension of the iterative scheme with improved and more accurate solutions. We can balance time and spatial scales in the scheme with optimal amounts of iterative steps. Such ideas consider an extended iterative scheme, such that we achieve a simple implementation of the scheme and reduce the computational time overall.

6.4.2 Extension Problem 2: Decomposition Methods for Hyperbolic Equations

In this section, we present benchmark problems of time decomposition methods for hyperbolic equations.

Based on the motivation of simulating wave equations, which are hyperbolic equations, we contribute iterative splitting methods as accurate solver methods with respect to decoupled complicated differential equations.

The splitting methods deal with simpler differential equations, with respecting time and space scales of the physical operators, and we are motivated to save memory and computational resources.

While using multisteps or more iterations, the additional amount of work for the operator-splitting methods can be reduced by optimizing the balance between time partitions and number of iteration steps.

In our case, the classical splitting methods for hyperbolic equations are the alternating direction implicit (ADI) methods [52], [155], as well as the locally one-dimensional (LOD) methods [40], [148]. The methods are based on locally reduced equations, e.g., explicitly parts, and sweep implicitly over all equation parts. These methods are often not stable and accurate enough, while neglecting the physical coupling of each operator, and delicate for designing higher-order methods, see [52].

In a first step, we achieve the improvements with a fourth-order LOD method and in a next step, we consider an iterative splitting scheme as a method to solve wave propagation problems. We achieve stability and consider higher-order results, while the iterative splitting method is based on fixed-point iterations, which also can be considered for wave equations. We gain higher-order results and the scheme can be easily implemented, see [55], [78], and [80].

We stability and consistency of the methods with the transformation on an abstract first-order Cauchy problem is done in [80].

We are interested in the spatially dependent wave equation

$$
\begin{aligned}
\frac{\partial^2 u}{\partial t^2} &= \frac{\partial}{\partial x_1}\tilde{D}_1(x)\frac{\partial u}{\partial x_1} + \ldots + \frac{\partial}{\partial x_d}\tilde{D}_d(x)\frac{\partial u}{\partial x_d} \text{ , in } \Omega \times [0,T], \\
u(x,0) &= u_0(x), \text{ in } \Omega, \\
\partial_t u(x,0) &= u_1(x), \text{ in } \Omega, \\
u(x,t) &= u_2(x,t), \text{ on } \partial\Omega_D \times [0,T], \\
\partial_n u(x,t) &= 0, \text{ on } \partial\Omega_N \times [0,T],
\end{aligned}
\tag{6.115}
$$

where $x = (x_1, \ldots, x_d)^T \in \Omega$, the initial functions are given as $u_0, u_1 : \Omega \to \mathbb{R}^+$, and the function for the Dirichlet boundary is $u_2 : \Omega \times [0,T] \to \mathbb{R}^+$. The domain $\Omega \subset \mathbb{R}^d$ is Lipschitz continuous and convex and the time interval is $[0,T] \subset \mathbb{R}^+$. The boundary is given as $\partial\Omega = \partial\Omega_D \cup \partial\Omega_N$. For the spatially dependent functions $\tilde{D}_1(x), \ldots, \tilde{D}_d(x) : \mathbb{R}^d \to \mathbb{R}^+$, we assume polynomial functions that map onto the positive real axis.

For constant wave parameters, i.e., $D_1, \ldots, D_d \in \mathbb{R}^+$, we can derive the analytical solution given by

$$u(x_1, \ldots, x_d, t) = \sin(\frac{1}{\sqrt{D_1}}\pi x_1) \cdot \ldots \cdot \sin(\frac{1}{\sqrt{D_d}}\pi x_d) \cdot \cos(\sqrt{d}\pi t). \quad (6.116)$$

We compute a reference solution for the nonconstant wave parameters with sufficiently fine grids and small time steps, see [182].

We now discuss noniterative and iterative splitting methods for two-dimensional and three-dimensional wave equations.

6.4.2.1 First Experiment: Elastic Wave Propagation with Noniterative Splitting Methods

In the following, we present simulations of test examples of the wave equation solved by noniterative splitting methods.

The noniterative splitting method in time discretized form is given by

$$
\begin{aligned}
\tilde{u} \quad - \quad & 2u^n + u^{n-1} \\
= \quad & \tau_n^2 D_1 \left(\eta \frac{\partial^2 \tilde{u}}{\partial x^2} + (1 - 2\eta)\frac{\partial^2 u^n}{\partial x^2} + \eta \frac{\partial^2 u^{n-1}}{\partial x^2} \right) \qquad (6.117) \\
+ \quad & \tau_n^2 D_2 \frac{\partial^2 u^n}{\partial y^2},
\end{aligned}
$$

$$
\begin{aligned}
u^{n+1} \quad - \quad & 2u^n + u^{n-1} \\
= \quad & \tau_n^2 D_1 \left(\eta \frac{\partial^2 \tilde{u}}{\partial x^2} + (1 - 2\eta)\frac{\partial^2 u^n}{\partial x^2} + \eta \frac{\partial^2 u^{n-1}}{\partial x^2} \right) \qquad (6.118) \\
+ \quad & \tau_n^2 D_2 \left(\eta \frac{\partial^2 u^{n+1}}{\partial y^2} + (1 - 2\eta)\frac{\partial^2 u^n}{\partial y^2} + \eta \frac{\partial^2 u^{n-1}}{\partial y^2} \right),
\end{aligned}
$$

where $\tau_n = t^{n+1} - t^n$ is the time step and the time discretization is of the second order.

First Test Example: Wave Equation with Constant Coefficients

The application of different boundary conditions are important for the splitting method used in the test example.

Dirichlet Boundary Condition
Our example is two dimensional and we can derive an analytical solution.

$$
\begin{aligned}
\partial_{tt} u \quad &= \quad D_1 \partial_{xx} u + D_2 \partial_{yy} u, \text{ in } \Omega \times [0, T], \qquad (6.119) \\
u(x, y, 0) \quad &= \quad u_{\text{exact}}(x, y, 0), \text{ in } \Omega, \\
\partial_t u(x, y, 0) \quad &= \quad 0, \text{ on } \Omega, \\
u(x, y, t) \quad &= \quad u_{\text{exact}}(x, y, t), \text{ on } \partial\Omega \times (0, T), \qquad (6.120)
\end{aligned}
$$

where $\Omega = [0, 1] \times [0, 1]$, $D_1 = 1$, $D_2 = 0.5$.

The analytical solution is given by

$$u_{\text{exact}}(x,y,t) = \sin(\frac{1}{\sqrt{D_1}}\pi x)\sin(\frac{1}{\sqrt{D_2}}\pi y)\cos(\sqrt{2}\,\pi t). \quad (6.121)$$

The discretization is given by an implicit time discretization and a finite difference method for the space discretization.

Thus, we have for the second-order discretization in space

$$
\begin{aligned}
Au(t) \;&= D_1\partial_{xx}u(t) \\
&\approx D_1\,\frac{u(x+\Delta x,y,t) - 2u(x,y,t) + u(x-\Delta x,y,t)}{\Delta x^2}, \quad (6.122) \\
Bu(t) \;&= D_2\partial_{yy}u(t) \\
&\approx D_2\,\frac{u(x,y+\Delta y,t) - 2u(x,y,t) + u(x,y-\Delta y,t)}{\Delta y^2}, \quad (6.123)
\end{aligned}
$$

where D_1, D_2 are positive constants and the second-order time discretization

$$\partial_{tt}u(t^n) \approx \frac{u(t^n + \tau_n) - 2u(t^n) + u(t^n - \tau_n)}{\tau_n^2}. \quad (6.124)$$

The implicit discretization is

$$
\begin{aligned}
u(t^{n+1}) - 2u(t^n) + u(t^{n-1}) \quad & \quad (6.125)\\
= \tau_n^2(A+B)(\eta u(t^{n+1}) + (1-2\eta)u(t^n) + \eta u(t^{n-1})).
\end{aligned}
$$

For the approximation error, we choose the L_1-norm; the L_2- and L_∞-norms are also possible.

The error in the L_1-norm is given by

$$err_{L_1} := \sum_{i,j=1,\ldots,m} V_{i,j}\,|u(x_i,y_j,t^n) - u_{\text{exact}}(x_i,y_j,t^n)|, \quad (6.126)$$

where $u(x_i,y_j,t^n)$ is the numerical solution and $u_{\text{exact}}(x_i,y_j,t^n)$ is the analytical solution. $V_{i,j} = \Delta x\,\Delta y$ is the mass, with Δx and Δy equidistant grid steps in the x and y dimensions. m is the number of grid nodes in the x and y dimensions.

In our test example, we choose $D_1 = D_2 = 1$ (a Dirichlet boundary) and set our model domain to be a rectangle $\Omega = [0,1] \times [0,1]$.

We discretize with $\Delta x = 1/16$, $\Delta y = 1/16$ and $\tau_n = 1/32$, and choose our parameter η between $0 \le \eta \le 0.5$.

The experimental results from finite differences, the classical operator splitting method, and the LOD method are shown in the following Tables 6.36, 6.37, 6.38, and Figure 6.12.

TABLE 6.36: Numerical results for the finite difference method (second-order finite differences in time and space) and a Dirichlet boundary.

η	err_{L_1}	u_{exact}	u_{num}
0.0	0.0014	-0.2663	-0.2697
0.1	0.0030	-0.2663	-0.2738
0.3	0.0063	-0.2663	-0.2820
0.5	0.0096	-0.2663	-0.2901

TABLE 6.37: Numerical results for the classical operator splitting method (second-order ADI method) and Dirichlet boundary.

η	err_{L_1}	u_{exact}	u_{num}
0.0	0.0014	-0.2663	-0.2697
0.1	0.0030	-0.2663	-0.2738
0.3	0.0063	-0.2663	-0.2820
0.5	0.0096	-0.2663	-0.2901

TABLE 6.38: Numerical results for the LOD method and Dirichlet boundary.

η	err_{L_1}	err_{L_∞}
0.0	0.0014	0.0034
0.1	0.0031	0.0077
0.3	0.0065	0.0161
0.5	0.0099	0.0245

Remark 6.15. In these experiments, we compare three different methods based on the ADI method. The splitting methods have the same quality of error reduction, while the most accurate method is the LOD method, which is a fourth-order method for $\eta = \frac{1}{12}$. The splitting methods are more effective methods.

Neumann Boundary Condition

Our next example is two dimensional, as we saw in the first example, but

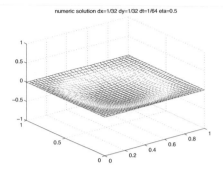

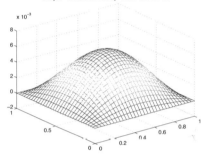

FIGURE 6.12: Numerical resolution of the wave equation: numerical approximation (top figure) and error functions (bottom figure) for the Dirichlet boundary ($\Delta x = \Delta y = 1/32$, $\tau_n = 1/64$, $D_1 = 1$, $D_2 = 1$, coupled version).

relative to Neumann boundary conditions.

$$
\begin{aligned}
\partial_{tt}u &= D_1\partial_{xx}u + D_2\partial_{yy}u, \text{ in } \Omega \times (0,T), & (6.127)\\
u(x,y,0) &= u_{\text{exact}}(x,y,0), \text{ in } \Omega,\\
\partial_t u(x,y,0) &= 0, \text{ in } \Omega,\\
\frac{\partial u(x,y,t)}{\partial n} &= \frac{\partial u_{\text{exact}}(x,y,t)}{\partial n}, \text{ on } \partial\Omega \times (0,T), & (6.128)
\end{aligned}
$$

where $\Omega = [0,1] \times [0,1]$, $D_1 = 1$, $D_2 = 0.5$ and we have equidistant time steps τ_n.

The analytical solution is given by

$$
u_{\text{exact}}(x,y,t) = \sin(\frac{1}{\sqrt{D_1}}\pi x)\sin(\frac{1}{\sqrt{D_2}}\pi y)\cos(\sqrt{2}\,\pi t). \quad (6.129)
$$

The experimental results of the finite difference method and classical operator splitting method are shown in the following Tables 6.39, 6.40, and Figure 6.13.

TABLE 6.39: Numerical results for the finite difference method (finite differences in time and space of second order) and a Neumann boundary.

η	err_{L_1}	u_{exact}	u_{num}
0.0	0.0014	-0.2663	-0.2697
0.1	0.0030	-0.2663	-0.2738
0.3	0.0063	-0.2663	-0.2820
0.5	0.0096	-0.2663	-0.2901
0.7	0.0128	-0.2663	-0.2981
0.9	0.0160	-0.2663	-0.3060
1.0	0.0176	-0.2663	-0.3100

TABLE 6.40: Numerical results for the classical operator splitting method (ADI method of second order) and a Neumann boundary.

η	err_{L_1}	u_{exact}	u_{num}
0.0	0.0014	-0.2663	-0.2697
0.1	0.0030	-0.2663	-0.2738
0.3	0.0063	-0.2663	-0.2820
0.5	0.0096	-0.2663	-0.2901

Remark 6.16. The Neumann boundary conditions are more delicate because of the underlying boundary conditions. Thus, we need additional work to split the boundary conditions into the spatial dimensions, which are applied in the ADI and LOD methods. The LOD method can be improved to a fourth-order method for $\eta = \frac{1}{12}$.

Second Test Example: Spatially Dependent Test Example

In this experiment, we apply our method to a spatially dependent problem.

For simplification and application of the numerical scheme, we consider only a nonconservative form of our wave equation.

We can also deal with conservative forms by adding convection terms, see the discretization schemes of conservative wave equations [44].

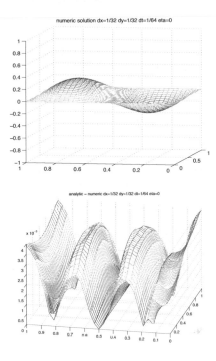

FIGURE 6.13: Numerical resolution of the wave equation: numerical approximation (top figure) and error functions (bottom figure) for Neumann boundary (right) ($\Delta x = \Delta y = 1/32$, $\tau_n = 1/64$, $D_1 = 1$, $D_2 = 0.5$, coupled version).

The underlying equation is given by

$$
\begin{aligned}
\partial_{tt}u &= D_1(x,y)\partial_{xx}u + D_2(x,y)\partial_{yy}u, \text{ in } \Omega \times (0,T), \quad (6.130)\\
u(x,y,0) &= u_0(x,y), \text{ in } \Omega,\\
\partial_t u(x,y,0) &= u_1(x,y), \text{ in } \Omega,\\
u(x,y,t) &= u_2(x,y,t), \text{ on } \partial\Omega \times (0,T),
\end{aligned}
$$

where $D_1(x,y) = 0.1x + 0.01y + 0.01$, $D_2(x,y) = 0.01x + 0.1y + 0.1$.

To compare the numerical results, we cannot use an analytical solution, which is why we compute a reference solution in an initial step. The reference solution is obtained with a finite difference scheme, with fine time and space steps.

When choosing the time steps, it is important to consider the CFL condition, which in this case is based on the spatial coefficients.

Remark 6.17. We have assumed the following CFL condition:

$$\tau_n < 0.5 \frac{\min(\Delta x, \Delta y)}{\max_{x,y \in \Omega}(D_1(x,y), D_2(x,y))}, \tag{6.131}$$

where Δx and Δy are equidistant spatial steps in the x and y dimensions.

For the test example, we define our model domain as a rectangle $\Omega = [0,1] \times [0,1]$.

The reference solution is obtained by executing the finite difference method and setting $\Delta x = 1/256, \Delta y = 1/256$ and a time step $\tau_n = 1/256 < 0.390625$.

The model domain is given by a rectangle with $\Delta x = 1/16$ and $\Delta y = 1/32$. The time steps are given by $\tau_n = 1/16$ and $0 \le \eta \le 0.5$.

The numerical results for Dirichlet boundary conditions are given in the following Tables 6.41, 6.42, 6.43, and Figure 6.14.

The results show the second-order accuracy and how similar results are obtained with the nonsplitting method (see Table 6.41) and classical splitting method (see Table 6.42). Therefore, the splitting method did not influence the numerical results and preserved the second-order character. The results can be improved by the LOD method (see Table 6.43) and it reaches fourth-order accuracy with the parameter $\eta = \frac{1}{12}$.

TABLE 6.41: Numerical results obtained by the finite difference method with spatially dependent parameters and Dirichlet boundary (error of reference solution).

η	err_{L_1}	u_{exact}	u_{num}
0.0	0.0032	-0.7251	-0.7154
0.1	0.0034	-0.7251	-0.7149
0.3	0.0037	-0.7251	-0.7139
0.5	0.0040	-0.7251	-0.7129

The numerical results for Neumann boundary conditions are given in the following Tables 6.44, 6.45, and Figure 6.15. We obtain the same results as shown in the Dirichlet case. The second-order accuracy of the classical splitting method (see 6.45) did not influence the second-order numerical results. The same improvement for the Neumann case to reach a higher order can be obtained with the LOD method, see [78].

TABLE 6.42: Numerical results obtained by the classical operator splitting method with spatially dependent parameters and Dirichlet boundary (error of reference solution).

η	err_{L_1}	u_{exact}	u_{num}
0.0	0.0032	-0.7251	-0.7154
0.1	0.0034	-0.7251	-0.7149
0.3	0.0037	-0.7251	-0.7139
0.5	0.0040	-0.7251	-0.7129

TABLE 6.43: Numerical results obtained by the LOD method with spatially dependent parameters and Dirichlet boundary (error of reference solution).

η	err_{L_1}	u_{exact}	u_{num}
0.00	0.0032	-0.7251	-0.7154
0.1	0.7809e-003	-0.7251	-0.7226
0.122	0.6793e-003	-0.7251	-0.7242
0.3	0.0047	-0.7251	-0.7369
0.5	0.0100	-0.7251	-0.7512

TABLE 6.44: Numerical results obtained by the finite difference method with spatially dependent parameters and Neumann boundary (error of reference solution).

η	err_{L_1}	u_{exact}	u_{num}
0.0	0.0180	-0.7484	-0.7545
0.1	0.0182	-0.7484	-0.7532
0.3	0.0185	-0.7484	-0.7504
0.5	0.0190	-0.7484	-0.7477

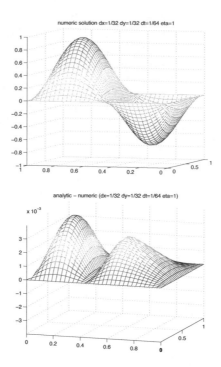

FIGURE 6.14: Dirichlet boundary condition: numerical solution (top figure) and error function (bottom figure) for the spatially dependent test example.

TABLE 6.45: Numerical results obtained by the classical operator splitting method with spatially dependent parameters and Neumann boundary (error of reference solution).

η	err_{L_1}	u_{exact}	u_{num}
0.0	0.0180	-0.7484	-0.7545
0.1	0.0182	-0.7484	-0.7532
0.3	0.0185	-0.7484	-0.7504
0.5	0.0190	-0.7484	-0.7477

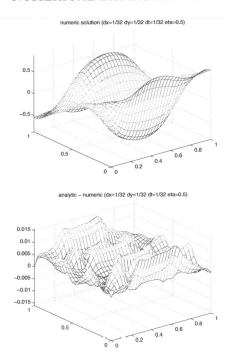

FIGURE 6.15: Neumann boundary condition: numerical solution (top figure) and error function (bottom figure) for the spatially dependent test example.

Remark 6.18. In our experiments, we have analyzed both the classical operator splitting and LOD methods and have shown that the LOD method yields more accurate values. At least the LOD method can be improved to a fourth-order method for $\eta = \frac{1}{12}$.

Remark 6.19. We have presented different time splitting methods for the spatially dependent case of the wave equation. The contributions of this chapter concern the boundary splitting and stiff operator treatment. For the boundary splitting method, we have discussed the theoretical background and the experiments show that the method is also stable for the stiff case. We have presented stable results even for the spatially dependent wave equation. The computational process benefits from decoupling the stiff and nonstiff operators into different equations, due to the different scales of the operators. The LOD method as a fourth-order method has the advantage of higher accuracy and can be used for such decoupling.

6.4.2.2   Second Experiment: Elastic Wave Propagation with Iterative Splitting Methods

In this section, we can improve the results with an iterative scheme that is embedded to the previously discussed LOD method. We discuss the benefits of the iterative modifications of the schemes and the balance between time steps and iterative steps.

We now consider iterative splitting methods to our two- and three-dimensional wave equations. The time discretization is of second order and our splitting method is given for 2D equations as

$$
\begin{aligned}
u^i(t^{n+1}) \quad - \quad & 2u^n + u^{n-1} \\
= \quad & \tau_n^2 D_1 \left(\eta \frac{\partial^2 u^i(t^{n+1})}{\partial x^2} + (1 - 2\eta) \frac{\partial^2 u^n}{\partial x^2} + \eta \frac{\partial^2 u^{n-1}}{\partial x^2} \right) \quad (6.132) \\
+ \quad & \tau_n^2 D_2 \left(\eta \frac{\partial^2 u^{i-1}(t^{n+1})}{\partial y^2} + (1 - 2\eta) \frac{\partial^2 u^n}{\partial y^2} + \eta \frac{\partial^2 u^{n-1}}{\partial y^2} \right),
\end{aligned}
$$

$$
\begin{aligned}
u^{i+1}(t^{n+1}) \quad - \quad & 2u^n + u^{n-1} \\
= \quad & \tau_n^2 D_1 \left(\eta \frac{\partial^2 u^i(t^{n+1})}{\partial x^2} + (1 - 2\eta) \frac{\partial^2 u^n}{\partial x^2} + \eta \frac{\partial^2 u^{n-1}}{\partial x^2} \right) \quad (6.133) \\
+ \quad & \tau_n^2 D_2 \left(\eta \frac{\partial^2 u^{i+1}(t^{n+1})}{\partial y^2} + (1 - 2\eta) \frac{\partial^2 u^n}{\partial y^2} + \eta \frac{\partial^2 u^{n-1}}{\partial y^2} \right),
\end{aligned}
$$

where $i = 1, 3, \ldots 2m + 1$ and the previous solutions are given as u^{n-1} and u^n, the starting solution $u^{i-1}(t^{n+1})$ is given as

$$
u^{i-1}(t^{n+1}) \quad - \quad 2u^n + u^{n-1} = \tau_n^2 \left(D_1 \frac{\partial^2 u^n}{\partial x^2} + D_2 \frac{\partial^2 u^n}{\partial y^2} \right).
$$

where for $\eta \in [0, 0.5]$ we have a second-order time discretization scheme, which is unconditionally stable for $\eta \in (0.25, 0.5)$, see [85], and [148].

The error is calculated as

$$
\mathrm{err}_{L_1} := \sum_{i=1}^{p} \Delta x_1 \cdot \ldots \cdot \Delta x_d \, |u(x_{1,i}, \ldots, x_{d,i}, t^n) - u_{\mathrm{exact}}(x_{1,i}, \ldots, x_{d,i}, t^n)|,
$$

$$(6.134)$$

where p is the number of grid points, u_{exact} is the analytical solution, and u is the numerical solution; $\Delta x_1, \ldots, \Delta x_d$ are equidistant spatial steps in each dimension.

In the following, we present simulations of test examples of the wave equation with iterative splitting methods.

First Test Example: Wave Equation with Constant Coefficients in Two Dimensions

In the first example, we compare the classical splitting method (ADI method) and iterative splitting methods for a nonstiff case and a stiff case.

We compare the nonstiff case for operators A and B.

We begin with $D_1 = 1$, $D_2 = \frac{1}{4}$, $\Delta x = \frac{1}{16}$, $\Delta y = \frac{1}{32}$, 2 iterations per time step, and $\Omega = [0,1]^2$, $t \in [0, 3 \cdot (1/\sqrt{2})]$.

We discretize with the temporal and spatial derivatives using second-order finite difference methods.

The results are given in Table 6.46 for the classical method (ADI method) and in Table 6.47 for the iterative splitting method.

TABLE 6.46: Classical operator splitting method; $D_1 = 1$, $D_2 = \frac{1}{4}$, $\Delta x = \frac{1}{16}$, $\Delta y = \frac{1}{32}$, and $\Omega = [0,1]^2$, $t \in [0, 3 \cdot (1/\sqrt{2})]$.

η	crr_{L_1}	crr_{L_1}	crr_{L_1}
tsteps	44	45	46
0	0.0235	2.3944e-004	4.8803e-007
0.0100	2.0405e-004	5.3070e-007	7.9358e-008
0.0800	4.1689e-005	4.1885e-005	4.2068e-005
0.0833	4.5958e-005	4.5970e-005	4.5982e-005
0.1000	7.0404e-005	6.9235e-005	6.8149e-005
0.2000	3.2520e-004	3.0765e-004	2.9168e-004

TABLE 6.47: Iterative splitting method; $D_1 = 1$, $D_2 = \frac{1}{4}$, $\Delta x = \frac{1}{16}$, $\Delta y = \frac{1}{32}$, two iterations per time step, starting solution with an explicit time step, and $\Omega = [0,1]^2$, $t \in [0, 3 \cdot (1/\sqrt{2})]$.

η	err_{L_1}	err_{L_1}	err_{L_1}
tsteps	44	45	46
0	0.0230	2.3977e-004	4.9793e-007
0.0100	1.4329e-004	1.2758e-006	7.9315e-008
0.0800	4.1602e-005	4.1805e-005	4.1994e-005
0.0833	4.5858e-005	4.5879e-005	4.5898e-005
0.1000	7.0228e-005	6.9074e-005	6.8003e-005
0.2000	3.2375e-004	3.0636e-004	2.9052e-004

In the next computations, we compare the stiff case for operators A and B.

We deal with $D_1 = \frac{1}{1}$, $D_2 = \frac{1}{100}$, $\Delta x = \frac{1}{16}$, $\Delta y = \frac{1}{32}$, and $\Omega = [0,1]^2 \times [0, 3 \cdot (1/\sqrt{2})]$. We discretize with the time and space derivatives using second-order finite difference methods.

The results are given in Table 6.48 for the classical method (ADI method) and in Table 6.49 for the iterative splitting method.

Remark 6.20. The iterative splitting method is in both cases more accurate

TABLE 6.48: Classical operator splitting method; $D_1 = \frac{1}{100}$, $D_2 = \frac{1}{100}$, $\Delta x = \frac{1}{16}$, $\Delta y = \frac{1}{32}$, and $\Omega = [0,1]^2$, $t \in [0, 3 \cdot (1/\sqrt{2})]$.

η	err_{L_1}	err_{L_1}	err_{L_1}	err_{L_1}	err_{L_1}	err_{L_1}
tsteps	3	5	6	7	15	100
0.0	41.0072	0.1205	9.2092e-004	0.0122	0.1131	0.1475
0.01	29.0629	0.0626	5.5714e-004	0.0212	0.1171	0.1476
0.05	6.9535	0.0100	0.0491	0.0764	0.1336	0.1479
0.08	0.9550	0.0984	0.1226	0.1331	0.1465	0.1482
0.09	0.1081	0.1409	0.1515	0.1539	0.1508	0.1483
0.1	0.3441	0.1869	0.1819	0.1755	0.1552	0.1484
0.2	0.7180	0.6259	0.4960	0.4052	0.2005	0.1494
0.3	0.2758	0.7949	0.7126	0.6027	0.2479	0.1504
0.4	0.0491	0.7443	0.7948	0.7316	0.2963	0.1513
0.5	0.0124	0.6005	0.7754	0.7905	0.3448	0.1523

TABLE 6.49: Iterative splitting method; $D_1 = \frac{1}{100}$, $D_2 = \frac{1}{100}$, $\Delta x = \frac{1}{16}$, $\Delta y = \frac{1}{32}$, starting solution with an explicit time step, and $\Omega = [0,1]^2$ $t \in [0, 3 \cdot (1/\sqrt{2})]$.

η	err_{L_1}	err_{L_1}	err_{L_1}	err_{L_1}	err_{L_1}	err_{L_1}
tsteps	3	5	6	7	15	100
# iter	4	8	10	12	28	198
0.0	11.5666	0.0732	6.1405e-004	0.0088	0.0982	0.1446
0.01	9.0126	0.0379	3.6395e-004	0.0151	0.1017	0.1447
0.05	3.3371	0.0047	0.0315	0.0540	0.1161	0.1450
0.08	1.4134	0.0520	0.0785	0.0935	0.1272	0.1453
0.09	0.9923	0.0756	0.0970	0.1081	0.1310	0.1454
0.1	0.6465	0.1019	0.1166	0.1231	0.1348	0.1455
0.2	0.7139	0.4086	0.3352	0.2877	0.1737	0.1465
0.3	0.7919	0.6541	0.5332	0.4466	0.2141	0.1474
0.4	0.6591	0.7963	0.6792	0.5787	0.2552	0.1484
0.5	0.5231	0.8573	0.7734	0.6791	0.2961	0.1493

and the best results are given for the discretization parameter $\eta \approx 0.01$. Two iteration steps are sufficient for obtaining second-order results. For the stiff case, more iteration steps are important so as to obtain the same results as for the nonstiff case.

Second Test Example: Wave Equation with Constant Coefficients in Three Dimensions

In the second example, we compare the iterative splitting methods for different starting solutions $u^{i-1}(t^{n+1})$.

We consider the explicit method first with the starting solution $u^{i-1,n+1}$, where

$$u^{i-1,n+1} - 2u^n + u^{n-1} = \tau_n^2 (D_1 \frac{\partial^2 u^n}{\partial x^2} + D_2 \frac{\partial^2 u^n}{\partial y^2}).$$

According to the second method, we calculate $u^{i-1,n+1}$ by applying the classical operator splitting method (ADI method), described in Section 6.4.2.1.

For our model equation, we apply the 3D wave equation:

$$\frac{\partial^2 u}{\partial t^2} = D_1 \frac{\partial^2 u}{\partial x^2} + D_2 \frac{\partial^2 u}{\partial y^2} + D_3 \frac{\partial^2 u}{\partial z^2}, \qquad (6.135)$$

where we have the exact initial and boundary conditions given by equation (6.116).

We use $D_1 = 1$, $D_2 = \frac{1}{10}$ and $D_3 = \frac{1}{100}$, $\Delta x = \Delta y = \Delta z = \frac{1}{8}$, 3 iterations per time step and $\Omega = [0, 1]^3$, $t \in [0, 6 \cdot (1/\sqrt{3})]$.

We discretize with time and space derivatives by applying second-order finite difference methods.

The results are given in Table 6.50 for the classical method (ADI method) and in Table 6.51 for the iterative splitting method.

TABLE 6.50: Iterative splitting method; $D_1 = 1$, $D_2 = \frac{1}{10}$ and $D_3 = \frac{1}{100}$, $\Delta x = \Delta y = \Delta z = \frac{1}{8}$, three iterations per time step, starting solution with an explicit time step, and $\Omega = [0, 1]^3$, $t \in [0, 6 \cdot (1/\sqrt{3})]$.

η	err_{L_1}	err_{L_1}	err_{L_1}	err_{L_1}	err_{L_1}
tsteps	8	9	10	11	12
0	169.4361	8.4256	0.5055	0.2807	0.3530
0.2000	0.0875	0.1315	0.1750	0.2232	0.2631
0.3000	0.1151	0.0431	0.0473	0.0745	0.1084
0.4000	0.3501	0.2055	0.0988	0.0438	0.0454
0.4500	0.4308	0.3002	0.1719	0.0844	0.0402
0.5000	0.4758	0.3792	0.2510	0.1402	0.0704

Remark 6.21. The starting solution prestep combined with the ADI method is more accurate and gives the best results for the discretization parameter $\eta \approx 0.01$. In the three-dimensional case, one more iteration step is needed than for the two-dimensional case. In comparison to the standard LOD schemes, we have an additional parameter of choosing the iterative steps. Such a new parameter balances between the time- and iterative steps and we can achieve larger time steps.

TABLE 6.51: Iterative splitting method; $D_1 = 1$, $D_2 = \frac{1}{10}$ and $D_3 = \frac{1}{100}$, $\Delta x = \Delta y = \Delta z = \frac{1}{8}$, three iterations per time step, starting solution with the classical splitting method (ADI) and $\Omega = [0, 1]^3$, $t \in [0 , 5 \cdot (1/\sqrt{3})]$.

η	err_{L_1}	err_{L_1}	err_{L_1}	err_{L_1}	err_{L_1}
tsteps	6	7	8	9	10
0	79.0827	10.1338	0.2707	0.1981	0.2791
0.2000	0.1692	0.2658	0.3374	0.3861	0.4181
0.3000	0.0311	0.0589	0.1182	0.1983	0.2692
0.4000	0.1503	0.0400	0.0389	0.0793	0.1322
0.4500	0.2340	0.0832	0.0316	0.0502	0.0930
0.5000	0.3101	0.1401	0.0441	0.0349	0.0645

6.4.3   Extension Problem 3: Nonlinear Partial Differential Equations

In this section, we discuss the application to nonlinear partial differential equations. The iterative splitting scheme is applied as a linearization scheme and we obtain linear partial differential equations. These linear equation systems can be solved with standard discretization methods, see [83].

The benefit of the iterative splitting scheme is to embed a linearization scheme and operate additionally as a decomposition scheme. Such multifunctional schemes reduce the computational time, see [146].

6.4.3.1   First Experiment: Burgers Equation

We deal with a 2D example for which we can derive an analytical solution.

$$\partial_t u = -u\partial_x u - u\partial_y u + \mu(\partial_{xx}u + \partial_{yy}u) + f(x,y,t), \quad (6.136)$$
$$\text{in } \Omega \times [0,T],$$

$$u(x,y,0) = u_{\text{analy}}(x,y,0), \text{ in } \Omega, \quad (6.137)$$

$$u(x,y,t) = u_{\text{analy}}(x,y,t) \text{ on } \partial\Omega \times [0,T], \quad (6.138)$$

where $\Omega = [0,1] \times [0,1]$, $T = 1.25$, and $\mu \in \mathbb{R}^+$ is the constant viscosity.

The analytical solution is given by

$$u_{\text{analy}}(x,y,t) = (1 + \exp(\tfrac{x+y-t}{2\mu}))^{-1}, \quad (6.139)$$

where $f(x,y,t) = 0$.

The operators are given by:

$$A(u)u = -u\partial_x u - u\partial_y u, \text{ hence} \quad (6.140)$$

$$A(u) = -u\partial_x - u\partial_y \text{ (nonlinear operator)},$$

$$Bu = \mu(\partial_{xx}u + \partial_{yy}u) + f(x,y,t) \text{ (linear operator)}. \quad (6.141)$$

We apply the iterative scheme with Equation (6.136) to the first part and obtain

$$A(u_{i-1})u_i = -u_{i-i}\partial_x u_i - u_{i-i}\partial_y u_i , \text{ and} \tag{6.142}$$

$$Bu_{i-1} = \mu(\partial_{xx} + \partial_{yy})u_{i-1} + f, \tag{6.143}$$

and we obtain linear operators because u_{i-1} is known from the previous time step.

Furthermore, we apply the iterative scheme based on Equation (6.136) to the second part and obtain:

$$A(u_{i-1})u_i = -u_{i-i}\partial_x u_i - u_{i-i}\partial_y u_i , \text{ and} \tag{6.144}$$

$$Bu_{i+1} = \mu(\partial_{xx} + \partial_{yy})u_{i+1} + f, \tag{6.145}$$

and we also have linear operators.

For the spatial discretization, we apply finite difference methods of second order. Furthermore, for the time discretization, we apply the Crank-Nicholson method as a second-order method in time. So we obtain an algebraic system, which we can solve directly using our MATLAB® software tool.

We apply two iterations for the iterative scheme. The end rule is given by:

$$||u_i - u_{i-1}|| \leq \epsilon, \tag{6.146}$$

$$\text{with } i = 1, 2, \ldots, \tag{6.147}$$

where $\epsilon = 0.001$.
The maximal error at end time $t = T$ is given by

$$\text{err}_{\max} = |u_{\text{num}} - u_{\text{ana}}| = \max_{i=1}^{p} |u_{\text{num}}(x_i, t) - u_{\text{ana}}(x_i, t)|,$$

and the numerical convergence rate is given by

$$\rho = \log(\text{err}_{h/2}/\text{err}_h)/\log(0.5).$$

We have the following results, see Tables 6.52 and 6.53, for different steps in time and space and for different viscosities.

Figure 6.40 presents the profile of the 2D nonlinear Burgers equation.

Remark 6.22. In the examples, we have two different cases of μ, which smoothen our equation. In the first test, we use a very small $\mu = 0.05$, so that we have a dominant hyperbolic behavior, and as a result we have a loss in regularity and a sharp front. The iterative splitting method loses one order. In the second test, we have increased the smoothness, so we get a more parabolic behavior. We can demonstrate improved results with higher accuracy.

TABLE 6.52:   Numerical results for Burgers equation with viscosity $\mu = 0.05$, initial condition $u_0(t) = u_n$, and two iterations per time step.

$\Delta x = \Delta y$	Δt	err_{L_1}	$\mathrm{err}_{\max}$	ρ_{L_1}	$\rho_{\max}$
1/10	1/10	0.0549	0.1867		
1/20	1/10	0.0468	0.1599	0.2303	0.2234
1/40	1/10	0.0418	0.1431	0.1630	0.1608
1/10	1/20	0.0447	0.1626		
1/20	1/20	0.0331	0.1215	0.4353	0.4210
1/40	1/20	0.0262	0.0943	0.3352	0.3645
1/10	1/40	0.0405	0.1551		
1/20	1/40	0.0265	0.1040	0.6108	0.5768
1/40	1/40	0.0181	0.0695	0.5517	0.5804

TABLE 6.53:   Numerical results for Burgers equation with viscosity $\mu = 5$, initial condition $u_0(t) = u_n$, and two iterations per time step.

$\Delta x = \Delta y$	Δt	err_{L_1}	$\mathrm{err}_{\max}$	ρ_{L_1}	$\rho_{\max}$
1/10	1/10	$1.1168 \cdot 10^{-7}$	$2.4390 \cdot 10^{-7}$		
1/20	1/10	$8.2098 \cdot 10^{-8}$	$1.7163 \cdot 10^{-7}$	0.4439	0.5070
1/40	1/10	$6.4506 \cdot 10^{-8}$	$1.3360 \cdot 10^{-7}$	0.3479	0.3614
1/10	1/20	$3.8260 \cdot 10^{-8}$	$9.0093 \cdot 10^{-8}$		
1/20	1/20	$2.5713 \cdot 10^{-8}$	$5.6943 \cdot 10^{-8}$	0.5733	0.6619
1/40	1/20	$1.8738 \cdot 10^{-8}$	$4.0020 \cdot 10^{-8}$	0.4565	0.5088
1/10	1/40	$1.9609 \cdot 10^{-8}$	$4.9688 \cdot 10^{-8}$		
1/20	1/40	$1.1863 \cdot 10^{-8}$	$2.8510 \cdot 10^{-8}$	0.7250	0.8014
1/40	1/40	$7.8625 \cdot 10^{-9}$	$1.8191 \cdot 10^{-8}$	0.5934	0.6482

Remark 6.23. In the examples, we have chosen the nonlinear operator to be A and the linear operator to be B. Based on the error analysis, this makes sense because the error is dominated by the operator B, while operator A is assumed to be solved exactly in the differential equations. Operator B is important in the initialization process and should be estimated carefully. More importantly, if we deal with two nonlinear operators A and B, such an operator should be B, which is more bounded $||B(u)u|| \leq ||A(u)u||$ for $u \in U$, U is the solution space.

Often such an estimation of the operators can be seen at the beginning, e.g., operator B is A-bounded like in the case of a convection operator B and a diffusion operator A, see also [49].

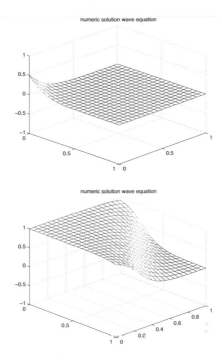

FIGURE 6.16: Burgers equation at initial time $t = 0.0$ (top figure) and end time $t = 1.25$ (bottom figure) for viscosity $\mu = 0.05$.

6.4.3.2   Second Experiment: Mixed Convection-Diffusion and Burgers Equation

We deal with a 2D example that is a mixture of convection-diffusion and the Burgers equation. We can derive an analytical solution.

$$
\begin{aligned}
\partial_t u &= -1/2u\partial_x u - 1/2u\partial_y u - 1/2\partial_x u - 1/2\partial_y u \\
&\quad + \mu(\partial_{xx}u + \partial_{yy}u) + f(x,y,t), \text{in } \Omega \times [0,T] \quad (6.148) \\
u(x,y,0) &= u_{\text{asym}}(x,y,0), \text{ in } \Omega \quad (6.149) \\
u(x,y,t) &= u_{\text{asym}}(x,y,t) \text{ on } \partial\Omega \times [0,T], \quad (6.150)
\end{aligned}
$$

where $\Omega = [0,1] \times [0,1]$, $T = 1.25$, and μ is the viscosity.

The analytical solution is given by

$$
u_{\text{asym}}(x,y,t) = (1 + \exp(\frac{x+y-t}{2\mu}))^{-1} + \exp(\frac{x+y-t}{2\mu}), \quad (6.151)
$$

where we compute $f(x,y,t)$ accordingly.

We split the convection-diffusion and Burgers equations. The operators are given by:

$$A(u)u \;=\; -1/2u\partial_x u - 1/2u\partial_y u + 1/2\mu(\partial_{xx}u + \partial_{yy}u), \text{ hence} \quad (6.152)$$
$$A(u) \;=\; 1/2(-u\partial_x - u\partial_y + \mu(\partial_{xx} + \partial_{yy})), \quad (6.153)$$

(Burgers term), and

$$Bu = -1/2\partial_x u - 1/2\partial_y u + 1/2\mu(\partial_{xx}u + \partial_{yy}u) + f(x,y,t), \quad (6.154)$$

(convection-diffusion term).

We apply the iterative scheme with Equation (6.148) to the first part and obtain:

$$A(u_{i-1})u_i \;=\; -1/2u_{i-1}\partial_x u_i - 1/2u_{i-1}\partial_y u_i \quad (6.155)$$
$$+1/2\mu(\partial_{xx}u_i + \partial_{yy}u_i),$$
$$Bu_{i-1} \;=\; 1/2(-\partial_x - \partial_y + \mu(\partial_{xx} + \partial_{yy}))u_{i-1}, \quad (6.156)$$

and we obtain linear operators, because u_{i-1} is known from the previous time step.

Furthermore, we apply the iterative scheme with Equation (6.148) to the second part and obtain:

$$A(u_{i-1})u_i \;=\; -1/2u_{i-1}\partial_x u_i - 1/2u_{i-1}\partial_y u_i \quad (6.157)$$
$$+1/2\mu(\partial_{xx}u_i + \partial_{yy}u_i),$$
$$Bu_{i+1} \;=\; 1/2(-\partial_x - \partial_y + \mu(\partial_{xx} + \partial_{yy}))u_{i+1}, \quad (6.158)$$

and we have linear operators.

For spatial discretization, we apply finite difference methods of the second order. Furthermore, for time discretization, we apply the Crank-Nicholson method as a second-order method in time. In this way, we obtain an algebraic system which we can solve directly using our MATLAB® software tool.

We apply two iterations to the iterative scheme. The end rule, see Equation (6.167), is given with $\epsilon = 0.001$.

We deal with different viscosities μ as well as different step sizes in time and space. We have the following results, see Tables 6.54 and 6.55.

Figure 6.17 presents the profile of the 2D linear and nonlinear convection-diffusion equation.

Remark 6.24. In the examples, we deal with more iteration steps to obtain higher-order convergence results. In the first test, we have four iterative steps but a smaller viscosity ($\mu = 0.5$), such that we can reach at least a second-order method. In the second test, we use a high viscosity of about $\mu = 5$ and

TABLE 6.54: Numerical results for mixed convection-diffusion and Burgers equation with viscosity $\mu = 0.5$, initial condition $u_0(t) = u_n$, and four iterations per time step.

$\Delta x = \Delta y$	Δt	err_{L_1}	$\mathrm{err}_{\max}$	ρ_{L_1}	$\rho_{\max}$
1/5	1/20	0.0137	0.0354		
1/10	1/20	0.0055	0.0139	1.3264	1.3499
1/20	1/20	0.0017	0.0043	1.6868	1.6900
1/40	1/20	$8.8839 \cdot 10^{-5}$	$3.8893 \cdot 10^{-4}$	4.2588	3.4663
1/5	1/40	0.0146	0.0377		
1/10	1/40	0.0064	0.0160	1.1984	1.2315
1/20	1/40	0.0026	0.0063	1.3004	1.3375
1/40	1/40	$8.2653 \cdot 10^{-4}$	0.0021	1.6478	1.6236

TABLE 6.55: Numerical results for mixed convection-diffusion and Burgers equation with viscosity $\mu = 5$, initial condition $u_0(t) = u_n$, and two iterations per time step.

$\Delta x = \Delta y$	Δt	err_{L_1}	$\mathrm{err}_{\max}$	ρ_{L_1}	$\rho_{\max}$
1/5	1/20	$1.3166 \cdot 10^{-5}$	$2.9819 \cdot 10^{-5}$		
1/10	1/20	$5.6944 \cdot 10^{-6}$	$1.3541 \cdot 10^{-5}$	1.2092	1.1389
1/20	1/20	$1.6986 \cdot 10^{-6}$	$4.5816 \cdot 10^{-6}$	1.7452	1.5634
1/40	1/20	$7.8145 \cdot 10^{-7}$	$2.0413 \cdot 10^{-6}$	1.1201	1.1663
1/5	1/40	$1.4425 \cdot 10^{-5}$	$3.2036 \cdot 10^{-5}$		
1/10	1/40	$7.2343 \cdot 10^{-6}$	$1.5762 \cdot 10^{-5}$	0.9957	1.0233
1/20	1/40	$3.0776 \cdot 10^{-6}$	$6.7999 \cdot 10^{-6}$	1.2330	1.2129
1/40	1/40	$9.8650 \cdot 10^{-7}$	$2.3352 \cdot 10^{-6}$	1.6414	1.5420

get a second-order result with two iteration steps. Here, we observe the loss of differentiability. To obtain the same results, we have to increase the number of iteration steps. So we were able to show an improvement in convergence order with respect to iteration steps.

6.4.3.3   Third Experiment: Momentum Equation (Molecular Flow)

We deal with the example of a momentum equation, which is used to model the viscous flow of fluid.

$$\partial_t \mathbf{u} = -\mathbf{u} \cdot \nabla \mathbf{u} + 2\mu \nabla (D(\mathbf{u}) + 1/3 \nabla \mathbf{u}) + \mathbf{f}(x, y, t),$$
$$\text{in } \Omega \times [0, T], \tag{6.159}$$

$$\mathbf{u}(x, y, 0) = \mathbf{g_1}(x, y), \text{ in } \Omega, \tag{6.160}$$

$$\mathbf{u}(x, y, t) = \mathbf{g_2}(x, y, t) \text{ on } \partial\Omega \times [0, T] \text{ (enclosed flow)}, \tag{6.161}$$

where $\mathbf{u} = (u_1, u_2)^t$ is the solution and $\Omega = [0, 1] \times [0, 1]$, $T = 1.25$, $\mu = 5$, $\mathbf{v} = (0.001, 0.001)^t$ are the parameters and I is the unit matrix.

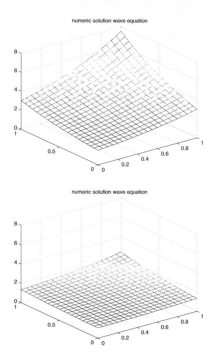

FIGURE 6.17: Mixed convection-diffusion and Burgers equation at initial time $t = 0.0$ (top figure) and end time $t = 1.25$ (bottom figure) for viscosity $\mu = 0.5$.

The nonlinear function $D(\mathbf{u}) = \mathbf{u} \cdot \mathbf{u} + \mathbf{v} \cdot \mathbf{u}$ is the viscous flow and $\mathbf{v}$ is a constant velocity.

We can derive the analytical solution with respect to the first two test examples with the functions:

$$u_1(x, y, t) = (1 + \exp(\frac{x + y - t}{2\mu}))^{-1} + \exp(\frac{x + y - t}{2\mu}), \quad (6.162)$$

$$u_2(x, y, t) = (1 + \exp(\frac{x + y - t}{2\mu}))^{-1} + \exp(\frac{x + y - t}{2\mu}). \quad (6.163)$$

For the splitting method our operators are given by:

$A(\mathbf{u})\mathbf{u} = -\mathbf{u}\nabla\mathbf{u} + 2\mu\nabla D(\mathbf{u})$ (nonlinear operator), and

$B\mathbf{u} = 2/3\mu\Delta\mathbf{u}$ (linear operator).

We first deal with the one-dimensional case,

$$\partial_t u = -u \cdot \partial_x u + 2\mu \partial_x (D(u) + 1/3 \partial_x u) \tag{6.164}$$
$$+ f(x,t), \text{ in } \Omega \times [0,T],$$
$$u(x,0) = g_1(x), \text{ in } \Omega, \tag{6.165}$$
$$u(x,t) = g_2(x,t) \text{ on } \partial\Omega \times [0,T] \text{ (enclosed flow)}, \tag{6.166}$$

where u is the solution and $\Omega = [0,1]$, $T = 1.25$, $\mu = 5$, and $v = 0.001$ are the parameters.

Then the operators are given by:

$A(u)u = -u\partial_x u + 2\mu\partial_x D(u)$ (nonlinear operator), and
$Bu = 2/3\mu\partial_{xx}u$ (linear operator).

For the spatial discretization, we apply finite difference methods of the second order. Furthermore, for time discretization, we apply the Crank-Nicholson method as a second-order method in time. In this way, we obtain an algebraic system that we can solve directly using our MATLAB® software tool.

For iterative splitting, we apply two iterations and the end rule, see Equation (6.167), is given with $\epsilon = 0.001$.

For iterative splitting used as a fixed-point scheme, we obtain the results displayed in Tables 6.56 to 6.58.

TABLE 6.56: Numerical results for 1D momentum equation with $\mu = 5$, $v = 0.001$, initial condition $u_0(t) = u_n$, and two iterations per time step.

Δx	Δt	err_{L_1}	$\text{err}_{\max}$	ρ_{L_1}	$\rho_{\max}$
1/10	1/20	0.0213	0.0495		
1/20	1/20	0.0203	0.0470	0.0689	0.0746
1/40	1/20	0.0198	0.0457	0.0401	0.0402
1/80	1/20	0.0195	0.0450	0.0216	0.0209
1/10	1/40	0.0134	0.0312		
1/20	1/40	0.0117	0.0271	0.1957	0.2009
1/40	1/40	0.0108	0.0249	0.1213	0.1211
1/80	1/40	0.0103	0.0238	0.0682	0.0674
1/10	1/80	0.0094	0.0217		
1/20	1/80	0.0073	0.0169	0.3591	0.3641
1/40	1/80	0.0062	0.0143	0.2451	0.2448
1/80	1/80	0.0056	0.0129	0.1478	0.1469

Figure 6.18 presents the profile of the 1D momentum equation. We have the following results for the 2D case, see Tables 6.59, 6.60, 6.61, and 6.62.

TABLE 6.57:   Numerical results for 1D momentum equation with $\mu = 5$, $v = 0.001$, initial condition $u_0(t) = u_n$, two iterations per time step, and $K = 1$ using the Newton iterative method.

Δx	Δt	err_{L_1}	$\mathrm{err}_{\max}$	ρ_{L_1}	$\rho_{\max}$
1/10	1/20	0.0180	0.0435		
1/20	1/20	0.0120	0.0276	0.5867	0.6550
1/40	1/20	0.0095	0.0227	0.3311	0.2870
1/80	1/20	0.0085	0.0208	0.1706	0.1231
1/10	1/40	0.0172	0.0459		
1/20	1/40	0.0125	0.0305	0.4652	0.5884
1/40	1/40	0.0108	0.0253	0.2366	0.2698
1/80	1/40	0.0097	0.0235	0.1191	0.1111
1/10	1/80	0.0166	0.0475		
1/20	1/80	0.0132	0.0338	0.3327	0.4917
1/40	1/80	0.0119	0.0280	0.1640	0.2734
1/80	1/80	0.0112	0.0265	0.0802	0.0779

TABLE 6.58:   Numerical results for 1D momentum equation with $\mu = 50$, $v = 0.1$, initial condition $u_0(t) = u_n$, and two iterations per time step.

Δx	Δt	err_{L_1}	$\mathrm{err}_{\max}$	ρ_{L_1}	$\rho_{\max}$
1/10	1/20	$2.7352 \cdot 10^{-6}$	$6.4129 \cdot 10^{-6}$		
1/20	1/20	$2.3320 \cdot 10^{-6}$	$5.4284 \cdot 10^{-6}$	0.2301	0.2404
1/40	1/20	$2.1144 \cdot 10^{-6}$	$4.9247 \cdot 10^{-6}$	0.1413	0.1405
1/80	1/20	$2.0021 \cdot 10^{-6}$	$4.6614 \cdot 10^{-6}$	0.0787	0.0793
1/10	1/40	$2.1711 \cdot 10^{-6}$	$5.2875 \cdot 10^{-6}$		
1/20	1/40	$1.7001 \cdot 10^{-6}$	$4.1292 \cdot 10^{-6}$	0.3528	0.3567
1/40	1/40	$1.4388 \cdot 10^{-6}$	$3.4979 \cdot 10^{-6}$	0.2408	0.2394
1/80	1/40	$1.3023 \cdot 10^{-6}$	$3.1694 \cdot 10^{-6}$	0.1438	0.1423
1/10	1/80	$1.6788 \cdot 10^{-6}$	$4.1163 \cdot 10^{-6}$		
1/20	1/80	$1.1870 \cdot 10^{-6}$	$2.9138 \cdot 10^{-6}$	0.5001	0.4984
1/40	1/80	$9.1123 \cdot 10^{-7}$	$2.2535 \cdot 10^{-6}$	0.3814	0.3707
1/80	1/80	$7.6585 \cdot 10^{-7}$	$1.9025 \cdot 10^{-6}$	0.2507	0.2443

Figure 6.19 presents the profile of the 2D momentum equation.

For Newton's method, we apply two iterations for the outer iterative scheme and three iterations for the inner iterative scheme, i.e., for Newton's method.

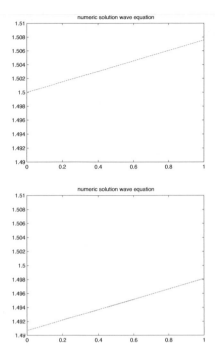

FIGURE 6.18: 1D momentum equation at initial time $t = 0.0$ (top figure) and end time $t = 1.25$ (bottom figure) for $\mu = 5$ and $v = 0.001$.

TABLE 6.59: Numerical results for 2D momentum equation with $\mu = 2$, $v = (1,1)^t$, initial condition $u_0(t) = u_n$, and two iterations per time step (first component).

Δx $= \Delta y$	Δt	err_{L_1} 1st c.	$\mathrm{err}_{\max}$ 1st c.	ρ_{L_1} 1st c.	$\rho_{\max}$ 1st c.
1/5	1/20	0.0027	0.0112		
1/10	1/20	0.0016	0.0039	0.7425	1.5230
1/20	1/20	0.0007	0.0022	1.2712	0.8597
1/5	1/40	0.0045	0.0148		
1/10	1/40	0.0032	0.0088	0.5124	0.7497
1/20	1/40	0.0014	0.0034	1.1693	1.3764
1/5	1/80	0.0136	0.0425		
1/10	1/80	0.0080	0.0241	0.7679	0.8197
1/20	1/80	0.0039	0.0113	1.0166	1.0872

The end rule is given by:

$$\|u_{1,i} - u_{1,i-1}\| \leq \epsilon_1, \tag{6.167}$$

$$\|u_{2,i} - u_{2,i-1}\| \leq \epsilon_2, \tag{6.168}$$

$$\text{with } i = 1, 2, \ldots, \tag{6.169}$$

TABLE 6.60: Numerical results for 2D momentum equation with $\mu = 2$, $v = (1,1)^t$, initial condition $u_0(t) = u_n$, and two iterations per time step (second component).

Δx $= \Delta y$	Δt	err_{L_1} 2nd c.	$\text{err}_{\max}$ 2nd c.	ρ_{L_1} 2nd c.	$\rho_{\max}$ 2nd c.
1/5	1/20	0.0145	0.0321		
1/10	1/20	0.0033	0.0072	2.1526	2.1519
1/20	1/20	0.0021	0.0042	0.6391	0.7967
1/5	1/40	0.0288	0.0601		
1/10	1/40	0.0125	0.0239	1.2012	1.3341
1/20	1/40	0.0029	0.0054	2.1263	2.1325
1/5	1/80	0.0493	0.1111		
1/10	1/80	0.0278	0.0572	0.8285	0.9579
1/20	1/80	0.0115	0.0231	1.2746	1.3058

TABLE 6.61: Numerical results for 2D momentum equation for first component with $\mu = 50$, $v = (100, 0.01)^t$, initial condition $u_0(t) = u_n$, and two iterations per time step.

Δx $= \Delta y$	Δt	err_{L_1} 1st c.	$\text{err}_{\max}$ 1st c.	ρ_{L_1} 1st c.	$\rho_{\max}$ 1st c.
1/5	1/20	$1.5438 \cdot 10^{-5}$	$3.4309 \cdot 10^{-5}$		
1/10	1/20	$4.9141 \cdot 10^{-6}$	$1.0522 \cdot 10^{-5}$	1.6515	1.7052
1/20	1/20	$1.5506 \cdot 10^{-6}$	$2.9160 \cdot 10^{-6}$	1.6641	1.8513
1/5	1/40	$2.8839 \cdot 10^{-5}$	$5.5444 \cdot 10^{-5}$		
1/10	1/40	$1.3790 \cdot 10^{-5}$	$2.3806 \cdot 10^{-5}$	1.0645	1.2197
1/20	1/40	$3.8495 \cdot 10^{-6}$	$6.8075 \cdot 10^{-6}$	1.8408	1.8061
1/5	1/80	$3.1295 \cdot 10^{-5}$	$5.5073 \cdot 10^{-5}$		
1/10	1/80	$1.7722 \cdot 10^{-5}$	$2.6822 \cdot 10^{-5}$	0.8204	1.0379
1/20	1/80	$7.6640 \cdot 10^{-6}$	$1.1356 \cdot 10^{-5}$	1.2094	1.2400

where $\epsilon_1 = 0.0001$ and $\epsilon_2 = 0.001$.

For the Newton operator splitting method, we obtain the following functional matrices for the one-dimensional case,

$$DF(u) = (4\mu - 1)\partial_x u, \tag{6.170}$$

TABLE 6.62: Numerical results for 2D momentum equation for second component with $\mu = 50$, $v = (100, 0.01)^t$, initial condition $u_0(t) = u_n$, and two iterations per time step.

Δx $= \Delta y$	Δt	err$_{L_1}$ 2nd c.	err$_{max}$ 2nd c.	ρ_{L_1} 2nd c.	ρ_{max} 2nd c.
1/5	1/20	$4.3543 \cdot 10^{-5}$	$1.4944 \cdot 10^{-4}$		
1/10	1/20	$3.3673 \cdot 10^{-5}$	$7.9483 \cdot 10^{-5}$	0.3708	0.9109
1/20	1/20	$2.6026 \cdot 10^{-5}$	$5.8697 \cdot 10^{-5}$	0.3717	0.4374
1/5	1/40	$3.4961 \cdot 10^{-5}$	$2.2384 \cdot 10^{-4}$		
1/10	1/40	$1.7944 \cdot 10^{-5}$	$8.9509 \cdot 10^{-5}$	0.9622	1.3224
1/20	1/40	$1.5956 \cdot 10^{-5}$	$3.6902 \cdot 10^{-5}$	0.1695	1.2783
1/5	1/80	$9.9887 \cdot 10^{-5}$	$3.3905 \cdot 10^{-4}$		
1/10	1/80	$3.5572 \cdot 10^{-5}$	$1.3625 \cdot 10^{-4}$	1.4896	1.3153
1/20	1/80	$1.0557 \cdot 10^{-5}$	$4.4096 \cdot 10^{-5}$	1.7525	1.6275

and

$$D(F(u)) = -\begin{pmatrix} \partial_{u_1} F_1(u) & \partial_{u_2} F_1(u) \\ \partial_{u_1} F_2(u) & \partial_{u_2} F_2(u) \end{pmatrix} \tag{6.171}$$

$$= -\begin{pmatrix} -\partial_x u_1 + 4\mu\partial_x u_1 & -\partial_x u_2 + 4\mu\partial_x u_2 \\ -\partial_y u_1 + 4\mu\partial_y u_1 & -\partial_y u_2 + 4\mu\partial_y u_2 \end{pmatrix}$$

$$= (4\mu - 1)\nabla u$$

for the two-dimensional case using

$$A(\mathbf{u})\mathbf{u} = -\mathbf{u}\nabla\mathbf{u} + 2\mu\nabla D(\mathbf{u}), \tag{6.172}$$

$$= -\begin{pmatrix} u_1\partial_x u_1 + u_2\partial_x u_2 \\ u_1\partial_y u_1 + u_2\partial_y u_2 \end{pmatrix}$$

$$+2\mu\begin{pmatrix} 2u_1\partial_x u_1 + 2u_2\partial_x u_2 + v_1\partial_x u_1 + v_2\partial_x u_2 \\ 2u_1\partial_y u_1 + 2u_2\partial_y u_2 + v_1\partial_y u_1 + v_2\partial_y u_2 \end{pmatrix} \tag{6.173}$$

Here, we do not need the linearization and apply the standard iterative splitting method.

We only linearize the first split step and therefore we can relax this step with the second linear split step. Therefore, we obtain stable methods, see [146].

Remark 6.25. In the more realistic examples of 1D and 2D momentum equations, we can also observe the stiffness problem, which we obtain with a more hyperbolic behavior. In 1D experiments, we deal with a more hyperbolic behavior and were able to obtain at least first-order convergence in two iterative steps. In 2D experiments, we obtain nearly second-order convergence results with two iterative steps, if we increase the parabolic behavior, e.g., larger μ and $\mathbf{v}$ values. For such methods, we have to balance the use of iterative steps, refinement in time and space with respect to the hyperbolicity of the equations. In the very least, we can obtain a second-order method in more than two iterative steps. So the stiffness influences the number of iterative steps.

6.4.4   Extension Problem 4: Coupled Equations

In this section, we discuss the application to coupled differential equations. The iterative splitting scheme is applied as a coupling scheme, but takes into account a fast computation of the matrix exponentials.

The benefit of the iterative splitting scheme is to embed a coupling and computational method, such that we can save additional CPU time.

6.4.5   First Example: Matrix Problem

We deal in the first experiment with an ODE and separate the complex operator in two simpler operators and consider the following matrix equation,

$$u'(t) = \begin{bmatrix} 1 & 2 \\ 3 & 0 \end{bmatrix} u, \quad u(0) = u_0 = \begin{pmatrix} 0 \\ 1 \end{pmatrix}, \tag{6.174}$$

the exact solution is

$$u(t) = 2(e^{3t} - e^{-2t})/5. \tag{6.175}$$

We split the matrix as,

$$\begin{bmatrix} 1 & 2 \\ 3 & 0 \end{bmatrix} = \begin{bmatrix} 1 & 1 \\ 1 & 0 \end{bmatrix} + \begin{bmatrix} 0 & 1 \\ 2 & 0 \end{bmatrix} \tag{6.176}$$

Figure 6.20 presents the numerical errors between the exact and the numerical solution.

Figure 6.21 presents the CPU time of the standard and the iterative splitting schemes.

6.4.6   Second Experiment: 10×10 Matrix

We deal in the second experiment with the 10×10 ODE system:

$$\partial_t u_1 = -\lambda_{1,1} u_1 + \lambda_{2,1} u_2 + \cdots + \lambda_{10,1} u_{10}, \tag{6.177}$$

$$\partial_t u_2 = \lambda_{1,2} u_1 - \lambda_{2,2}(t) u_2 + \cdots + \lambda_{10,2} u_{10}, \tag{6.178}$$

$$\vdots \tag{6.179}$$

$$\partial_t u_{10} = \lambda_{1,10} u_1 + \lambda_{2,10}(t) u_2 + \cdots - \lambda_{10,10} u_{10}, \tag{6.180}$$

$$u_1(0) = u_{1,0}, \ldots, u_{10}(0) = u_{10,0} \text{ (initial conditions)}, \tag{6.181}$$

where $\lambda_1(t) \in \mathbb{R}^+$ and $\lambda_2(t) \in \mathbb{R}^+$ are the decay factors and $u_{1,0}, \ldots, u_{10,0} \in \mathbb{R}^+$. We have the time interval $t \in [0, T]$.

We rewrite the Equation (6.177) in operator notation, concentrating on the following equations:

$$\partial_t u = A(t)u + B(t)u , \qquad (6.182)$$

where $u_1(0) = u_{10} = 1.0$, $u_2(0) = u_{20} = 1.0$ are the initial conditions, and where we have $\lambda_1(t) = t$ and $\lambda_2(t) = t^2$.

The operators are splitted in the following way

$$A = \begin{pmatrix} -0.01 & 0.01 & 0 & \cdots & & & & & & \\ 0.01 & -0.01 & 0 & \cdots & & & & & & \\ 0.01 & 0.01 & -0.02 & 0 & \cdots & & & & & \\ 0.01 & 0.01 & 0.01 & -0.03 & 0 & \cdots & & & & \\ \vdots & & & & & & & & & \\ 0.01 & 0.01 & 0.01 & 0.01 & 0.01 & 0.01 & 0.01 & 0.01 & -0.08 & 0 \\ 0.01 & 0.01 & 0.01 & 0.01 & 0.01 & 0.01 & 0.01 & 0.01 & 0 & -0.08 \end{pmatrix} , \qquad (6.183)$$

$$B = \begin{pmatrix} -0.08 & 0 & 0.01 & 0.01 & 0.01 & 0.01 & 0.01 & 0.01 & 0.01 & 0.01 \\ 0 & -0.08 & 0.01 & 0.01 & 0.01 & 0.01 & 0.01 & 0.01 & 0.01 & 0.01 \\ \vdots & & & & & & & & & \\ 0 & 0 & 0 & 0 & 0 & 0 & 0 & -0.02 & 0.01 & 0.01 \\ 0 & 0 & 0 & 0 & 0 & 0 & 0 & 0 & -0.01 & 0.01 \\ 0 & 0 & 0 & 0 & 0 & 0 & 0 & 0 & 0.01 & -0.01 \end{pmatrix} . \qquad (6.184)$$

Figure 6.22 presents the numerical errors between the exact and the numerical solution.

Figure 6.23 presents the CPU time of the standard and the iterative splitting schemes.

Remark 6.26. *The computational results show the benefit of the iterative schemes. We save computational time and achieve higher-order accuracy. The one-side and two-side schemes have the same results.*

6.4.7 Third Example: Commutator Problem

We assume to have a large norm of the commutator $[A, B]$ and deal with the matrix equation,

$$u'(t) = \begin{bmatrix} 10 & 1 \\ 1 & 10 \end{bmatrix} u, \quad u(0) = u_0 = \begin{pmatrix} 1 \\ 1 \end{pmatrix}, \qquad (6.185)$$

In the following, we discuss different splitting ideas.

Version 1:
We split the matrix as,

$$\begin{bmatrix} 10 & 1 \\ 1 & 10 \end{bmatrix} = \begin{bmatrix} 9 & 0 \\ 1 & 1 \end{bmatrix} + \begin{bmatrix} 1 & 1 \\ 0 & 9 \end{bmatrix} \qquad (6.186)$$

while $|||[A, B]||| \geq \max\{||A||, ||B||\}$.

Figure 6.24 presents the numerical errors between the exact and the numerical solution.

Figure 6.25 presents the CPU time of the standard and the iterative splitting schemes.

Further for the one-side scheme, we obtain more improved results for the following splitting.

Version 2:

$$\begin{bmatrix} 10 & 1 \\ 1 & 10 \end{bmatrix} = \begin{bmatrix} 9 & 0 \\ 1 & 9 \end{bmatrix} + \begin{bmatrix} 1 & 1 \\ 0 & 1 \end{bmatrix} \tag{6.187}$$

The Figure 6.26 present the numerical errors between the exact and the numerical solution.

A more delicate problem is given for the stiff matrices.

Version 3:

$$\begin{bmatrix} 10^4 & 1 \\ 1 & 10^4 \end{bmatrix} = \begin{bmatrix} 10^4 - 1 & 0 \\ 1 & 10^4 - 1 \end{bmatrix} + \begin{bmatrix} 1 & 1 \\ 0 & 1 \end{bmatrix} \tag{6.188}$$

Figure 6.27 presents the numerical errors between the exact and the numerical solution.

Figure 6.28 presents the CPU time of the standard and the iterative splitting schemes.

Remark 6.27. *The iterative schemes with fast computations of the exponential matrices have a speedup. The constant CPU time of the iterative schemes shows its benefits instead of the expensive standard schemes. Also for stiff problems with multi-iterative steps, we reach the same results for the standard A-B or Strang-Splitting schemes.*

6.4.8 Two-Phase Example

The equation is given as:

$$\partial_t c_1 + \nabla \cdot \mathbf{F} c_1 = g(-c_1 + c_{1,im}) - \lambda_1 c_1, \text{ in } \Omega \times [0, t], \tag{6.189}$$

$$\partial_t c_2 + \nabla \cdot \mathbf{F} c_2 = g(-c_2 + c_{2,im}) + \lambda_1 c_1 - \lambda_2 c_2, \tag{6.190}$$

$$\text{in } \Omega \times [0, t],$$

$$\mathbf{F} = \mathbf{v} - D\nabla, \tag{6.191}$$

$$\partial_t c_{1,im} = g(c_1 - c_{1,im}) - \lambda_1 c_{1,im}, \text{ in } \Omega \times [0, t], \tag{6.192}$$

$$\partial_t c_{2,im} = g(c_2 - c_{2,im}) + \lambda_1 c_{1,im} - \lambda_2 c_{2,im}, \text{ in } \Omega \times [0, t], \tag{6.193}$$

$$c_1(\mathbf{x}, t) = c_{1,0}(\mathbf{x}), c_2(\mathbf{x}, t) = c_{2,0}(\mathbf{x}), \text{ on } \Omega, \tag{6.194}$$

$$c_1(\mathbf{x}, t) = c_{1,1}(\mathbf{x}, t), c_2(\mathbf{x}, t) = c_{2,1}(\mathbf{x}, t), \text{ on } \partial\Omega \times [0, t], \tag{6.195}$$

$$c_{1,im}(\mathbf{x}, t) = 0, c_{2,im}(\mathbf{x}, t) = 0, \text{ on } \Omega, \tag{6.196}$$

$$c_{1,im}(\mathbf{x}, t) = 0, c_{2,im}(\mathbf{x}, t) = 0, \text{ on } \partial\Omega \times [0, t], \tag{6.197}$$

In the following equations, we deal with the semidiscretized equation given with the matrices:

We assume two species $m = 2$ and have the following two operators for the splitting method:

$$A = \frac{D}{\Delta x^2} \cdot \begin{pmatrix} -2 & 1 & & & \\ 1 & -2 & 1 & & \\ & \ddots & \ddots & \ddots & \\ & & 1 & -2 & 1 \\ & & & 1 & -2 \end{pmatrix} \tag{6.198}$$

$$+ \frac{v}{\Delta x} \cdot \begin{pmatrix} 1 & & & \\ -1 & 1 & & \\ & \ddots & \ddots & \\ & & -1 & 1 \\ & & & -1 & 1 \end{pmatrix} \in \mathbb{R}^{I \times I} \tag{6.199}$$

where I is the number of spatial points.

$$\Lambda_1 = \begin{pmatrix} \lambda_1 & 0 & & & \\ 0 & \lambda_1 & 0 & & \\ & \ddots & \ddots & \ddots & \\ & & 0 & \lambda_1 & 0 \\ & & & 0 & \lambda_1 \end{pmatrix} \in \mathbb{R}^{I \times I} \tag{6.200}$$

$$\Lambda_2 \; = \; \begin{pmatrix} \lambda_2 & 0 & & & \\ 0 & \lambda_2 & 0 & & \\ & \ddots & \ddots & \ddots & \\ & & 0 & \lambda_2 & 0 \\ & & & 0 & \lambda_2 \end{pmatrix} \; \in \; \mathbb{R}^{I \times I} \qquad (6.201)$$

$$G \; = \; \begin{pmatrix} g & 0 & & & \\ 0 & g & 0 & & \\ & \ddots & \ddots & \ddots & \\ & & 0 & g & 0 \\ & & & 0 & g \end{pmatrix} \; \in \; \mathbb{R}^{I \times I} \qquad (6.202)$$

We obtain the two matrices:

$$A_1 \; = \; \begin{pmatrix} A & 0 & 0 & 0 \\ 0 & A & 0 & 0 \\ 0 & 0 & 0 & 0 \\ 0 & 0 & 0 & 0 \end{pmatrix} \; \in \; \mathbb{R}^{4I \times 4I} \qquad (6.203)$$

$$\tilde{A}_2 \; = \; \begin{pmatrix} -\Lambda_1 & 0 & 0 & 0 \\ \Lambda_1 & -\Lambda_2 & 0 & 0 \\ 0 & 0 & -\Lambda_1 & 0 \\ 0 & 0 & \Lambda_1 & -\Lambda_2 \end{pmatrix} \; \in \; \mathbb{R}^{4I \times 4I} \qquad (6.204)$$

$$\tilde{A}_3 \; = \; \begin{pmatrix} -G & 0 & G & 0 \\ 0 & -G & 0 & G \\ G & 0 & -G & 0 \\ 0 & G & 0 & -G \end{pmatrix} \; \in \; \mathbb{R}^{4I \times 4I} \qquad (6.205)$$

For the operator A_1 and $A_2 = \tilde{A}_2 + \tilde{A}_3$ we apply the splitting method.

Figure 6.29 presents the numerical errors between the exact and the numerical solution.

Remark 6.28. *For all iterative schemes, we can reach faster results for the iterative schemes with fast computations of the exponential matrices standard schemes. With four to five iterative steps we obtain more accurate results than we did for the expensive standard schemes. With one-side iterative schemes we reach the best convergence results.*

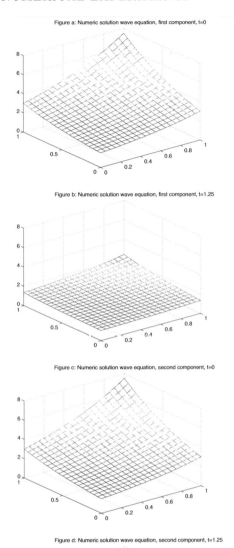

Figure a: Numeric solution wave equation, first component, t=0

Figure b: Numeric solution wave equation, first component, t=1.25

Figure c: Numeric solution wave equation, second component, t=0

Figure d: Numeric solution wave equation, second component, t=1.25

FIGURE 6.19: 2D momentum equation at initial time $t = 0.0$ (figure a,c) and end time $t = 1.25$ (figure b,d) for $\mu = 0.5$ and $v = (1,1)^t$ for the first and second components of the numerical solution.

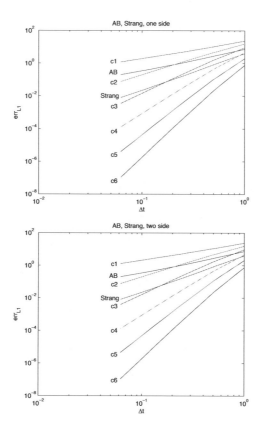

FIGURE 6.20: Numerical errors of the standard splitting scheme and the iterative schemes with $1, \ldots, 6$ iterative steps (top: comparison to the one side scheme, bottom: comparision to the two side scheme).

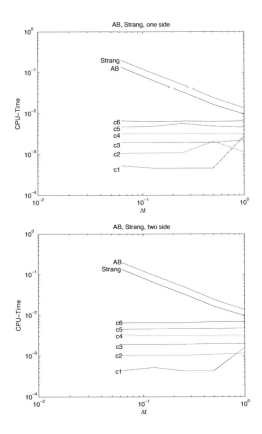

FIGURE 6.21: CPU time of the standard splitting scheme and the iterative schemes with $1, \ldots, 6$ iterative steps (top: comparison to the one side scheme, bottom: comparsion to the two side scheme).

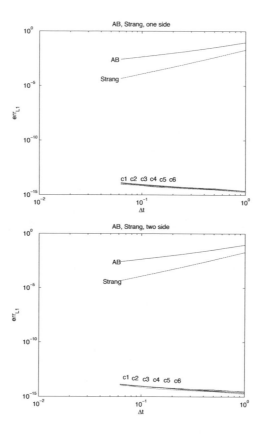

FIGURE 6.22: Numerical errors of the standard splitting scheme and the iterative schemes with $1, \ldots, 6$ iterative steps (top: comparison to the one side scheme, bottom: comparison to the two side scheme).

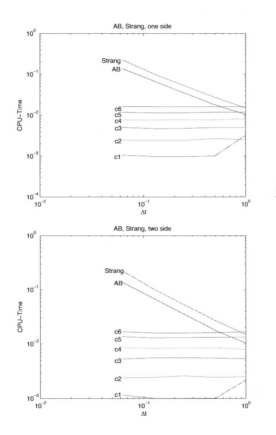

FIGURE 6.23: CPU time of the standard splitting scheme and the iterative schemes with $1, \ldots, 6$ iterative steps (top: comparison to the one side scheme, bottom: comparison to the two side scheme).

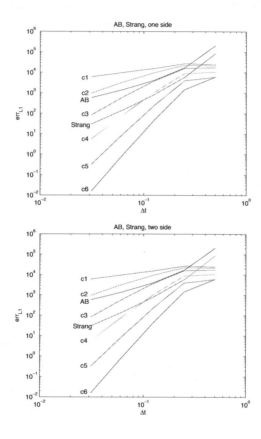

FIGURE 6.24: Numerical errors of the standard splitting scheme and the iterative schemes with $1, \ldots, 6$ iterative steps (top: comparison to the one side scheme, bottom: comparison to the two side scheme).

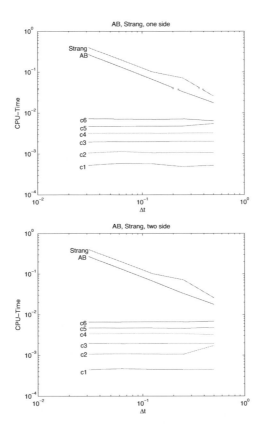

FIGURE 6.25: CPU time of the standard splitting scheme and the iterative schemes with $1, \ldots, 6$ iterative steps (top: comparison to the one side scheme, bottom: comparison to the two side scheme).

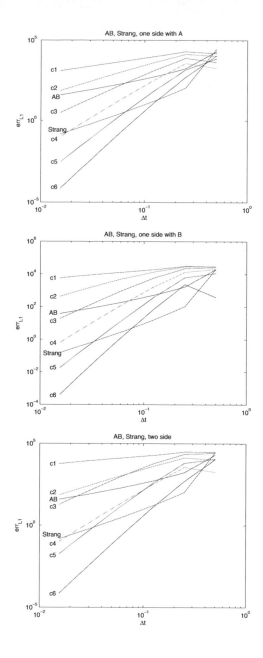

FIGURE 6.26: Numerical errors of the standard splitting scheme and the iterative schemes based on one side to operator A or B and two side scheme (top: comparison to the one side with A scheme, middle: comparison to the one side with B scheme, bottom: comparison to the two side scheme).

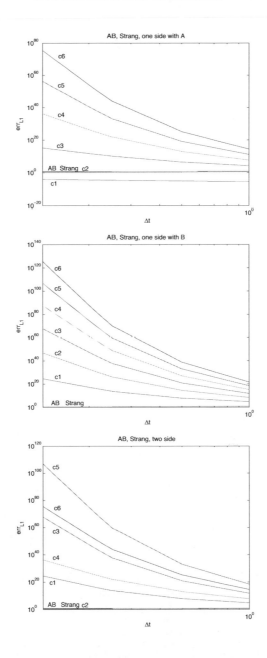

FIGURE 6.27: Numerical errors of the standard splitting scheme and the iterative schemes with $1, \ldots, 6$ iterative steps (top: comparison to the one side with A scheme, middle: comparison to the one side with B scheme, bottom: comparison to the two side scheme).

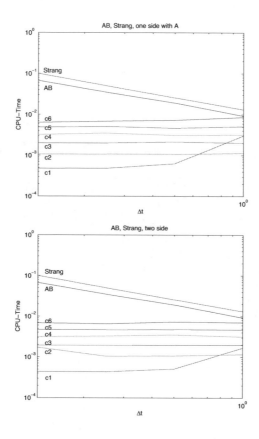

FIGURE 6.28: CPU time of the standard splitting scheme and the iterative schemes with $1, \dots, 6$ iterative steps (top: comparison to the one side with A scheme, bottom: comparison to the two side scheme).

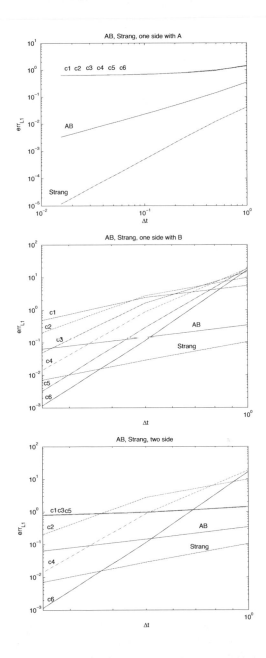

FIGURE 6.29: Numerical errors of the standard splitting scheme and the iterative schemes with $1, \ldots, 6$ iterative steps (top: comparison to the one side with A scheme, middle: comparison to the one side with B scheme, bottom: comparison to the two side scheme).

6.5 Real-Life Applications

In this section we deal with the real-life applications that were briefly introduced in Chapter 1. We discuss the efficiency and accuracy of our proposed splitting methods with respect to underlying discretization and solver methods for more realistic problems. While the time and spatial discretization schemes are more delicate and time consuming, we only consider second order approaches. For such schemes, it is sufficient to deal with lower-order splitting schemes, e.g., we propose two iterative steps in the iterative splitting method or AB-splitting schemes. For wave propagation problems, we consider LOD schemes of maximal fourth-order accuracy, such that we could improve this scheme with embedded iterative splitting schemes.

We have the following thesis:

- Reduction of computational time with more efficient schemes.

- Higher-order accuracy, while separating to simpler solvable parts.

- Simple implementation and a simpler extension of model equations with the coupling idea.

More or less we deal with multiphysics problems, which contain different physical processes, see also [111]. Especially, we discuss:

- Waste disposal (transport processes).

- Elastics wave propagation (wave processes).

- CVD apparatus (transport and retardation processes)

- Complex flow phenomena (flow processes)

6.5.1 Waste Disposal: Transport and Reaction of Radioactive Contaminants

In the next subsections, we describe the 2D and 3D simulations of waste disposal. Here the underlying ideas are to simulate complicated transport regions in permeable and nonpermeable layers. The interest is to see the contamination at the surface of the overlying domain. The splitting idea is to decompose the transport and reaction part of the equations. We solve the transport equation with second-order time discretization (BDF2 or Crank-Nicolson) and the spatial discretization with second order finite volume methods, see [88]. For iterative splitting schemes, it was sufficient to apply two iterative steps, because of the barrier of the second-order schemes in the time discretization.

6.5.1.1 Two-Dimensional Model of Waste Disposal

We calculate some waste disposal scenarios that help us to get new conclusions about waste disposal in salt domes.

We consider a model based on a salt dome with a layer of overlying rock, with a permanent source of groundwater flow that becomes contaminated with radioactive waste. Based on our model, we calculate the transport and the reaction of these contaminants coupled with decay chains, see [63]. The simulation time is $10000[a]$ and we calculate the concentration of waste in the water that flows to the top of the overlying rock. With these dates we can determine if the waste disposal is safe. We present a two-dimensional test case with the data of our last project partner GRS in Braunschweig (Germany), cf. [60] and [61].

We have a model domain with the size of $6000[m] \times 150[m]$ consisting of four different layers with different permeabilities, see [60]. We compute the 26 components, where the retardation and reaction parameters are given in [60]. The parameters for the diffusion and dispersion tensor are given by:
$D = 1 \cdot 10^{-9}[m^2/s]$, $\alpha_L = 4.0$ $[m]$, $\alpha_T = 0.4$ $[m]$, $|v|_{max} = 6 \cdot 10^{-6}[m/s]$, $\rho = 2 \cdot 10^3$, $D_L = \alpha_L|v|$ and $D_T = \alpha_T|v|$, where the longitudinal dispersion length is ten times bigger than the transversal dispersion.

Groundwater is pooled in the domain from the right to the left boundary. The groundwater flows faster through the permeable layer than through the impermeable layers. Therefore, the groundwater flows from the right boundary to the middle half of the domain. It flows through the permeable layer down to the bottom of the domain and pools at the top left of the domain to an outflow at the left boundary. The flow field with its velocity is calculated with the program package $\mathbf{D^3F}$ (Distributed-Density-Driven-Flow software toolbox, see [58]), and is presented in Figure 6.30.

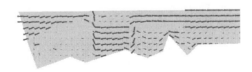

FIGURE 6.30: Flow field for a two-dimensional calculation.

In the middle of the bottom of the domain, the contaminants flow as if from a permanent source. With a stationary velocity field, the contaminants are computed with the software package **R³T** (Radioactive-Reaction-Retardation-Transport software toolbox, see [59]). The flow field transports the radioactive contaminants up to the top of the domain. The decay chain is presented with 28 components as follows,

$$Pu-244 \rightarrow Pu-240 \rightarrow U-236 \rightarrow Th-232 \rightarrow Ra-228$$

$$Cm-244 \rightarrow Pu-240$$

$$U-232$$

$$Pu-241 \rightarrow Am-241 \rightarrow Np-237 \rightarrow U-233 \rightarrow Th-229$$

$$Cm-246 \rightarrow Pu-242 \rightarrow U-238 \rightarrow U-234 \rightarrow Th-230 \rightarrow$$

$$Ra-226 \rightarrow Pb-210$$

$$Am-242 \rightarrow Pu-238 \rightarrow U-234$$

$$Am-243 \rightarrow Pu-239 \rightarrow U-235 \rightarrow Pa-231 \rightarrow Ac-227\,.$$

The retardation and decay factors are extremely varying over multiple decimal powers, see for example the neptunium sequence.

The decay chain is denoted by:
Am-241 $\rightarrow$ Np-237 $\rightarrow$ U-233 $\rightarrow$ Th-229.
The retardation factors are:
$R_{Am} = 200.2$, $R_{Np} = 2.2$, $R_U = 1.6$, $R_{Th} = 2000.2$.
The porosity is:
$\phi = 0.5$.
The half-life periods are:
$T_{1/2,Am} = 1.3639 \, 10^{10}$ [s], $T_{1/2,Np} = 6.7659 \, 10^{13}$ [s],
$T_{1/2,U} = 15.0239 \, 10^{12}$ [s], $T_{1/2,Th} = 2.4867 \, 10^{11}$ [s].

Here, we have to solve as accurately as possible the reaction equations, which are ODEs to reduce the splitting error of the A-B of iterative splitting schemes.

Further, we present the important concentration in this decay chain and applied embedded analytical solver methods to reduce computational time, see [70]. In Figure 6.31, the contaminant uranium isotope U-236 is presented after 100[a]. The movement of this isotope is less retarded by adsorption than other nuclides and has a very long half-life period. The concentration of U-236 in the model domain is illustrated in Figure 6.31 at two different time points in the modeling. The diffusion process spreads the contaminant throughout the left side of the domain. The impermeable layer is also contaminated. After a time period of 10000[a], the contaminant flows up to the top of the domain.

The calculations are performed on uniform grids. The convergence of these grids are confirmed with adaptive grid calculations, which confirmed the results of finer and smaller time steps, cf. Table 6.63. The calculations begin with explicit methods until the character of the equation is more diffusive; at this point we switch to the implicit methods and use large time steps. With

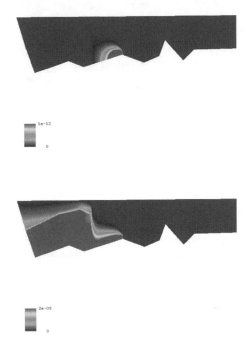

FIGURE 6.31: Concentration of U-236 at the time point $t = 100[a]$ and $t = 10000[a]$.

this procedure, we can remain within the limits of a mandatory maximum calculation time of one day.

TABLE 6.63: Computing the two-dimensional case.

Processors	Refinement	Number of elements	Number of time steps	Time for one time step	total time
30	uniform	75000	3800	5 sec.	5.5 h.
64	adaptive	350000	3800	14 sec.	14.5 h.

6.5.1.2 Three-Dimensional Model of Waste Disposal

In this example, we consider a three-dimensional model, because of the interest in the three-dimensional effects of contaminants in the flow of groundwater. We simulate approximately $10000[a]$ and focus on the important contaminants flowing furthest with a high concentration. We assume an

anisotropy domain of $6000[m] \times 2000[m] \times 1500[m]$ with different permeable layers. We have calculated 28 components as presented in the two-dimensional case. The parameters for the diffusion and dispersion tensor are given by:
$D = 1 \cdot 10^{-9} [m^2/s]$, $\alpha_L = 4.0\ [m]$, $\alpha_T = 0.4\ [m]$, $|v|_{max} = 6 \cdot 10^{-6} [m/s]$, $\rho = 2 \cdot 10^3$, $D_L = \alpha_L |v|$ and $D_T = \alpha_T |v|$, where the longitudinal dispersion length is ten times bigger than the transversal dispersion. The source is situated at the point $(4250.0, 2000.0, 1040.0)$ and the contaminants flow with a constant rate. The underlying velocity fields are calculated with $\mathbf{D^3F}$, and we added two sinks at the surface with the coordinates $(2000, 2100, 2073)$ and $(2500, 2000, 2073)$.

We simulate the transport and the reaction of the contaminants with our software package $\mathbf{R^3T}$. As a result, we can simulate the pumping rate of the sinks needed to pump out all of the contaminated groundwater. We present the velocity field in Figure 6.32. The groundwater flows from the right boundary to the middle of the domain. Due to the impermeable layers, the groundwater flows downward and pools in the middle part of the domain. Because of the influence of the salt dome, salt wells up with the groundwater and becomes incorporated into the lower middle part of the domain curls. These parts are interesting for 3D calculations, and because of the curls, the groundwater pools.

FIGURE 6.32: Flow field for a three-dimensional calculation.

We concentrate on the important component $U - 236$. This component is less retarded and is drawn into the sinks. In the top images of Figures 6.33, we show the initial concentration at time point $t = 100\ [a]$. We present the data in vertical cut planes. In the next picture, we present a cut plane through the source term. In the bottom of Figures 6.33, the concentration is presented at the end time point $t = 10000\ [a]$. The concentration has welled from the bottom up over the impermeable layer and into the sinks at the top of the domain.

At the beginning of the calculation, we use explicit discretization methods with respect to the convection-dominant case. After the initializing process, the contaminants are spread out with the diffusion process. We use the implicit methods with larger time steps and could also calculate within the mandatory time period even using a higher-order discretization method.

The computations are performed on a Linux PC-Cluster with 1.6 GHz Athlon processor.

In Table 6.64, we report the results of the computations. We begin with convergence results on uniform refined meshes. We confirm these results with adaptive refined meshes and obtain the same results with smaller time steps. We accomplished these calculations within the one-day mandatory computational time limit.

TABLE 6.64: Three-dimensional calculations of a realistic potential damage event.

Processors	Refinement	Number of elements	Number of time steps	Time for one time step	Total time
16	uniform	531264	3600	13.0 sec.	13.0 h
72	adaptive	580000	3600	18.5 sec.	18.5 h

Remark 6.29. Here the performance is obvious, based on the decoupling of the equation into convection-reaction and diffusion parts. We could accelerate the solution process with the most adapted discretization and solution methods. The resources are saved with direct ODE solvers for the reaction part and the possibility of large time steps for the implicit discretized diffusion part. For such problems a decoupling makes sense with regard to the parallelization of the underlying solution methods.

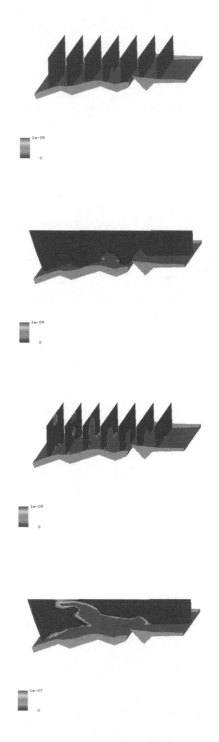

FIGURE 6.33: Concentration of U-236 at the time points $t = 100[a]$ and $t = 10000[a]$.

6.5.2 Elastic Wave Propagation

In these sections, we focus on decoupling the wave equations motivated by a realistic problem involving seismic sources and waves. The complicated wave propagation process is decomposed into simpler processes with respect to the directions of the propagation. We discretize and solve the simpler equations by more accurate higher-order methods, see [76].

We verify, with test examples, these decomposition methods for linear acoustic wave equations. Moreover, the delicate initialization of the right-hand side, while using a Dirac function, is presented with respect to the decomposition method. The benefit of spatial splitting methods, applied to multidimensional operators, is discussed in several examples.

6.5.2.1 Real-Life Application of Elastic Wave Propagation

In this application, we refer to our underlying model equations presented in Section 1.2.2. Based on these equations, we discuss their discretization and decomposition.

We focus on a LOD method, which can be improved to a higher-order splitting method. Such ideas are discussed in Subsection 6.4.2. The higher-order scheme for the real-life application problem has involved the approximations of time and space; we also focus on how the Dirac function is approximated. Based on the barrier of the fourth-order splitting scheme, we can improve such LOD methods with higher-order time discretization schemes, see also [33]. Further improvement to the splitting scheme can be done with an embedded iterative splitting scheme, such that we can balance the time discretization and splitting errors, which is discussed in Subsection 6.4.2.

During numerical testing we have observed a need to reduce the allowable time step in the case where the ratio of λ to μ becomes too large. This is likely to be due to the influence of the explicitly treated mixed derivative. For really high ratios (> 20), a reduction of 35% was necessary to avoid numerical instabilities. Here we discuss elastic wave propagation given as a system of linear wave equations, as introduced in Chapter 1.

6.5.2.2 Basic Numerical Methods

The standard explicit difference scheme with central differences was introduced in [5]. To save space, we first give an example in two dimensions.

We discretize uniformly in space and time on the unit square. We then get a grid Ω_h with grid points $x_j = jh, y_k = kh$, where $h > 0$ is the spatial grid size, and a time interval $[0, T]$ with the time points $t_n = n\tau_n$, where τ_n is the equidistant time step. The indices are denoted as $j \in \{0, 1, \ldots, J\}$, $k \in \{0, 1, \ldots, K\}$, and $n \in \{0, 1, \ldots, N\}$, with $J, K, N \in \mathbb{N}^+$. If we define the grid function by $U_{j,k}^n = U(x_j, y_k, t_n)$ with $U = (u, v)^T$ or $U = (u, v, w)$, the basic explicit scheme is given by

$$\rho \frac{U_{j,k}^{n+1} - 2U_{j,k}^n + U_{j,k}^{n-1}}{\tau_n^2} = \mathcal{M}_2 U_{j,k}^n + f_{j,k}^n, \tag{6.206}$$

where $\mathcal{M}_2$ is a difference operator:

$$\mathcal{M}_2 = \begin{pmatrix} (\lambda + 2\mu)\partial_x^2 + \mu\partial_y^2 & (\lambda + \mu)\partial_x^0\partial_y^0 v_{j,k} \\ (\lambda + \mu)\partial_x^0\partial_y^0 v_{j,k} & (\lambda + 2\mu)\partial_y^2 + \mu\partial_x^2 \end{pmatrix}, \tag{6.207}$$

and we use the standard difference operator notation:

$$\partial_x^+ v_{j,k} = \frac{1}{h}(v_{j+1,k} - v_{j,k}), \quad \partial_x^- v_{j,k} = \partial_x^+ v_{j-1,k}, \quad \partial_x^0 = \frac{1}{2}\left(\partial_x^+ + \partial_x^-\right),$$

and

$$\partial_x^2 = \partial_x^+ \partial_x^-.$$

$\mathcal{M}_2$ approximates $\mu\nabla^2 I + (\lambda + \mu)\nabla(\nabla\cdot)$ to the second order, where I is the identity matrix. This explicit scheme is stable for time steps satisfying

$$\tau_n < \frac{h}{\lambda + 3\mu}. \tag{6.208}$$

Replacing $\mathcal{M}_2$ by $\mathcal{M}_4$, we have a fourth-order difference operator given by

$$\mathcal{M}_4 = \begin{pmatrix} (\lambda + 2\mu)\left(1 - \frac{h^2}{12}\partial_x^2\right)\partial_x^2 + \mu\left(1 - \frac{h^2}{12}\partial_y^2\right)\partial_y^2 \\ (\lambda + \mu)\left(1 - \frac{h^2}{6}\partial_x^2\right)\partial_x^0\left(1 - \frac{h^2}{6}\partial_y^2\right)\partial_y^0 \\ (\lambda + \mu)\left(1 - \frac{h^2}{6}\partial_x^2\right)\partial_x^0\left(1 - \frac{h^2}{6}\partial_y^2\right)\partial_y^0 \\ (\lambda + 2\mu)\left(1 - \frac{h^2}{12}\partial_y^2\right)\partial_y^2 + \mu\left(1 - \frac{h^2}{12}\partial_x^2\right)\partial_x^2 \end{pmatrix}, \tag{6.209}$$

and using the modified equation approach [33], we obtain the explicit fourth-order scheme

$$\rho \frac{U_{j,k}^{n+1} - 2U_{j,k}^n + U_{j,k}^{n-1}}{\tau_n^2} = \mathcal{M}_4 U_{j,k}^n + f_{j,k}^n \tag{6.210}$$

$$+ \frac{\tau_n^2}{12}(\mathcal{M}_2 U_{j,k}^n + \mathcal{M}_2 f_{i,j}^n + \partial_{tt} f_{i,j}^n),$$

where $\mathcal{M}_2$ is a second-order approximation to the $\left(\mu\nabla^2 I + (\lambda + \mu)\nabla(\nabla\cdot)\right)^2$ operator, i.e., the right-hand side operator squared. As $\mathcal{M}_2$ only needs to be accurate to the second order, it has the same extent in space as $\mathcal{M}_4$, and no more grid points are used. The extra terms on the right-hand side are used to eliminate the second-order error in the approximation of the time derivative, and we obtain a fully fourth-order scheme.

In [148], the following implicit scheme for the scalar wave equation was introduced:

$$\rho \frac{U_{j,k}^{n+1} - 2U_{j,k}^n + U_{j,k}^{n-1}}{\tau_n^2} = \mathcal{M}_4 \left(\theta U_{j,k}^{n+1} + (1 - 2\theta)U_{j,k}^n + \theta U_{j,k}^{n-1} \right) \quad (6.211)$$

$$+ \theta f_{j,k}^{n+1} + (1 - 2\theta)f_{j,k}^n + \theta f_{j,k}^{n-1}.$$

When $\theta = 1/12$, the error of this scheme is the fourth order in time and space. For this θ value it is, however, only conditionally stable.

In order to make it competitive with the explicit scheme (6.210), we will provide operator split versions of the implicit scheme (6.211). The presence of the mixed derivative terms, that couple different coordinate directions, makes the creation of split versions complicated.

6.5.2.3 Fourth-Order Splitting Method

Here, we present a fourth-order splitting method based on the basic scheme (6.211). We split the operator $\mathcal{M}_4$ into three parts: $\mathcal{M}_{xx}$, $\mathcal{M}_{yy}$, and $\mathcal{M}_{xy}$, where we have

$$\mathcal{M}_{xx} = \begin{pmatrix} (\lambda + 2\mu)\left(1 - \frac{h^2}{12}\partial_x^2\right)\partial_x^2 & 0 \\ 0 & \mu\left(1 - \frac{h^2}{12}\partial_x^+\partial_x^-\right)\partial_x^2 \end{pmatrix},$$

$$\mathcal{M}_{yy} = \begin{pmatrix} \mu\left(1 - \frac{h^2}{12}\partial_y^2\right)\partial_y^2 & 0 \\ 0 & (\lambda + 2\mu)\left(1 - \frac{h^2}{12}\partial_y^2\right)\partial_y^2 \end{pmatrix},$$

$$\mathcal{M}_{xy} = \mathcal{M}_4 - \mathcal{M}_{xx} - \mathcal{M}_{yy}.$$

Our proposed splitting method has the following steps:

1. $\quad \rho \dfrac{U_{j,k}^* - 2U_{j,k}^n + U_{j,k}^{n-1}}{\tau_n^2} = \mathcal{M}_4 U_{j,k}^n \quad\quad\quad\quad (6.212)$

$$+ \theta f_{j,k}^{n+1} + (1 - 2\theta)f_{j,k}^n + \theta f_{j,k}^{n-1},$$

2. $\quad \rho \dfrac{U_{j,k}^{**} - 2U_{j,k}^*}{\tau_n^2} = \theta \mathcal{M}_{xx}\left(U_{j,k}^{**} - 2U_{j,k}^n + U_{j,k}^{n-1}\right) \quad (6.213)$

$$+ \frac{\theta}{2}\mathcal{M}_{xy}\left(U_{j,k}^* - 2U_{j,k}^n + U_{j,k}^{n-1}\right),$$

3. $\quad \rho \dfrac{U_{j,k}^{n+1} - 2U_{j,k}^{**}}{\tau_n^2} = \theta \mathcal{M}_{xx}\left(U_{j,k}^{n+1} - 2U_{j,k}^n + U_{j,k}^{n-1}\right) \quad (6.214)$

$$+ \frac{\theta}{2}\mathcal{M}_{xy}\left(U_{j,k}^{**} - 2U_{j,k}^n + U_{j,k}^{n-1}\right).$$

Here the first step is explicit, while the second and third steps treat the derivatives along the coordinate axes implicitly and the mixed derivatives explicitly. This is similar to how the mixed case is handled for parabolic problems, see [16].

6.5.2.4   Initial Values and Boundary Conditions

In order to start the time stepping scheme, we need to know the values at two earlier time levels. Starting at time $t = 0$, we know the value at level $n = 0$ as $U^0 = g_0$. The value at level $n = -1$ can be obtained by the Taylor expansion as follows:

$$U^{-1} = U^0 - \tau \partial_t U^0 + \frac{\tau_n^2}{2} \partial_{tt} U^0 - \frac{\tau^3}{6} \partial_{ttt} U^0 + \frac{\tau^4}{24} \partial_{tttt} U^0 + \mathcal{O}(\tau^5), \quad (6.215)$$

where we use

$$\partial_t U^0_{j,k} = g_{1\,j,k}, \quad (6.216)$$

$$\partial_{tt} U^0_{j,k} \approx \frac{1}{\rho} \left(\mathcal{M}_4 g_{0\,j,k} \right) + f_{j,k} \right), \quad (6.217)$$

$$\partial_{ttt} U^0_{j,k} \approx \frac{1}{\rho} \left(\mathcal{M}_4 g_{1\,j,k} \right) + \partial_t f^0_{\,j,k} \right), \quad (6.218)$$

$$\partial_{tttt} U^0_{\,j,k} \approx \frac{1}{\rho} \left(\mathcal{M}_2^2 g_{0\,j,k} \right) + \mathcal{M}_4 f^0_{\,j,k} + \partial_{tt} f^0_{\,j,k} \right), \quad (6.219)$$

and also for (6.218) and (6.219).

The approximation of our right-hand side is given as

$$\partial_t f^0_{\,j,k} \approx \frac{f^1_{\,j,k} - f^{-1}_{\,j,k}}{2\tau}, \quad (6.220)$$

$$\partial_{tt} f^0_{\,j,k} \approx \frac{f^1_{\,j,k} - 2 f^0_{\,j,k} + f^{-1}_{\,j,k}}{\tau_n^2}. \quad (6.221)$$

We are not considering the boundary value problem and thus we will not be concerned with constructing proper difference stencils at grid points near the boundaries of the computational domain. We have simply added a two-point-thick layer of extra grid points at the boundaries of the domain and assigned the correct analytical solution at all points in the layer for every time step.

Remark 6.30. For the Dirichlet boundary conditions, the splitting method, see (6.212)–(6.214), also conserves the conditions. For the three equations, i.e., for U^*, U^{**}, and U^{n+1}, we can use the same conditions.

For the Neumann boundary conditions and other boundary conditions of higher order, we also have to split the boundary conditions with respect to the split operators, see [166].

6.5.2.5   Test Example of the 2D Wave Equation

The first test example is a two-dimensional example. The splitting method is presented in Subsection 6.5.2.3 as well as the model equations.

We apply to the first test case a forcing function given as

$$
f = \quad \big(\sin(t-x)\sin(y) - 2\mu\sin(t-x)\sin(y)
$$
$$
-(\lambda+\mu)(\cos(x)\cos(t-y) + \sin(t-x)\sin(y)),
$$
$$
\sin(t-y)\sin(x) - 2Vs^2\sin(x)\sin(t-y)
$$
$$
-(\lambda+\mu)(\cos(t-x)\cos(y) + \sin(y)\sin(t-y))\big)^T, \quad (6.222)
$$

giving the analytical solution

$$
U^{\mathrm{true}} = \big(\sin(x-t)\sin(y), \ \sin(y-t)\sin(x)\big)^T. \tag{6.223}
$$

Using the splitting method, we solved Equation (1.2) on a domain $\Omega = [-1,1] \times [-1,1]$ and time interval $t \in [0,2]$. We used two sets of material parameters; for the first case ρ, λ, and μ were all equal to one, for the second case ρ and μ were one, and λ was set to 14. Solving on four different grids with a refinement factor of two in each direction between the successive grids, we obtained the results shown in Table 6.65. For all test examples, the equidistant time step is given as $\tau = 0.0063$. The errors are measured in the L_∞-norm defined as $\|U_{j,k}\| = \max\left(\max_{j,k}|u_{j,k}|, \max_{j,k}|v_{j,k}|\right)$. As can be

TABLE 6.65: Errors in max-norm for decreasing h and smooth analytical solution U^{true}. Convergence rate indicates fourth-order convergence for the split scheme.

Grid step h	Time $t=2$, $e_h = \mathrm{err}_{U,L_\infty} = \|U^n - U^{\mathrm{true}}\|_\infty$			
	case 1	$\log_2\left(\frac{e_{2h}}{e_h}\right)$	case 2	$\log_2\left(\frac{e_{2h}}{e_h}\right)$
0.05	1.7683e-07		2.5403e-07	
0.025	1.2220e-08	3.855	2.1104e-08	3.589
0.0125	7.9018e-10	3.951	1.4376e-09	3.876
0.00625	5.0013e-11	3.982	9.2727e-11	3.955

seen, we get the expected fourth-order convergence for problems with smooth solutions.

To check the influence of the splitting error $\mathcal{N}_{4,\theta}$ on the error, we solved the same problems using the nonsplit scheme (6.211). The results are shown in Table 6.66. The errors are only marginally smaller than for the split scheme.

6.5.2.6 Singular Forcing Terms

In seismology and acoustics, it is common to use spatial singular forcing terms that can look like

$$
f = F\delta(x)g(t), \tag{6.224}
$$

where F is a constant direction vector. A numeric method for Equation (1.2), see Chapter 1, needs to approximate the Dirac function correctly in order to achieve full convergence. Obviously, we cannot expect convergence close

TABLE 6.66: Errors in max-norm for decreasing h and smooth analytical solution U^{true} and using the nonsplit scheme. Comparing with Table 6.65 we see that the splitting error is very small for this case.

Grid step h	Time $t = 2$, $e_h = \text{err}_{U,L_\infty} = \|U^n - U^{\text{true}}\|_\infty$	
	case 1	case 2
0.05	1.6878e-07	2.4593e-07
0.025	1.1561e-08	2.0682e-08
0.0125	7.4757e-10	1.4205e-09
0.00625	4.8112e-11	9.2573e-11

to the source as the solution will be singular for two- and three-dimensional domains.

The analyses in [196] and [200] demonstrate that it is possible to derive regularized approximations of the Dirac function, that result in pointwise convergence of the solution away from the sources. Based on these analyses, we define one second- (δ_{h^2}) and one fourth- (δ_{h^4}) order regularized approximation of the one-dimensional Dirac function,

$$\delta_{h^2}(\tilde{x}) = \frac{1}{h} \begin{cases} 1 + \tilde{x}, & -h \leq \tilde{x} < 0, \\ 1 - \tilde{x}, & 0 \leq \tilde{x} < h, \\ 0, & \text{elsewhere,} \end{cases} \tag{6.225}$$

$$\delta_{h^4}(\tilde{x}) = \frac{1}{h} \begin{cases} 1 + \frac{11}{6}\tilde{x} + \frac{5}{6}\tilde{x}^2 + \frac{1}{6}\tilde{x}^3, & -2h \leq \tilde{x} < -h, \\ 1 + \frac{1}{2}\tilde{x} - \tilde{x}^2 - \frac{1}{2}\tilde{x}^3, & -h \leq \tilde{x} < 0, \\ 1 - \frac{1}{2}\tilde{x} - \tilde{x}^2 + \frac{1}{2}\tilde{x}^3, & 0 \leq \tilde{x} < h, \\ 1 - \frac{11}{6}\tilde{x} + \tilde{x}^2 - \frac{1}{6}\tilde{x}^3, & h \leq \tilde{x} < 2h, \\ 0, & \text{elsewhere,} \end{cases} \tag{6.226}$$

where $\tilde{x} = x/h$. The two- and three-dimensional Dirac functions are then approximated as $\delta_{h^{2,4}}(\tilde{x})\delta_{h^{2,4}}(\tilde{y})$ and $\delta_{h^{2,4}}(\tilde{x})\delta_{h^{2,4}}(\tilde{y})\delta_{h^{2,4}}(\tilde{z})$. The chosen time dependence was a smooth function given by

$$g(t) = \begin{cases} \exp(-1/(t(1-t))), & 0 \leq t < 1, \\ 0, & \text{elsewhere,} \end{cases} \tag{6.227}$$

which is C^∞. Using this forcing function, we can compute the analytical solution by integrating the Green's function given in [46]. The integration was done using numerical quadrature routines from MATLAB®. Figures 6.34 and 6.35 show examples of what the errors look like on a radius passing through the singular source at time $t = 0.8$ for different grid sizes h and the two approximations δ_{h^2} and δ_{h^4}.

We see that the error is smooth and converges a small distance away from the source. However, using δ_{h^2} limits the convergence to the second order, while using δ_{h^4} gives the full fourth-order convergence away from the singular source.

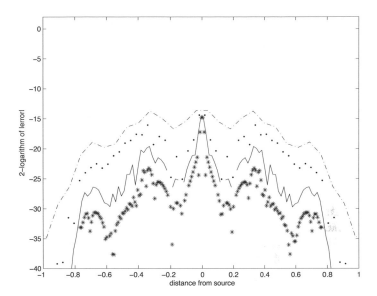

FIGURE 6.34: The 2-logarithm of the error along a line going through the source point for a point force located at $x = 0, y = 0$, and approximated in space by (6.226). Note that the error decays as $\mathcal{O}(h^4)$ away from the source, but not near it. The grid sizes were $h = 0.05\ (-\cdot)$, $0.025\ (\cdot)$, $0.0125\ (-)$, $0.00625\ (*)$. The numerical quadrature had an absolute error of approximately $10^{-11} \approx 2^{-36}$, so the error cannot be resolved beneath that limit.

When $t > 1$, the forcing goes to zero and the solution will be smooth everywhere. Table 6.67 shows the convergence behavior at time $t = 1.1$ for four different grids. Note that full convergence is achieved even if the lower order δ_{h^2} is used as an approximation for the Dirac function.

The convergence rate approaches 4 as we refine the grids, even though the solution was singular up to time $t = 1$.

We implemented the 2D case in MATLAB® on a single CPU. Further computations were also done on multiple CPUs with optimal speed.

The spatial grid steps are given as
$h = \Delta x = \Delta y = \Delta y \in \{0.05, 0.025, 0.0125, 0.00625, 0.003125\}$ and the time step is given as $\tau = 0.0063$. The material parameters for the elastic wave propagation equations are $\lambda = 1$, $\mu = 1$, and $\rho = 1$. The total number of grid points is about 26000.

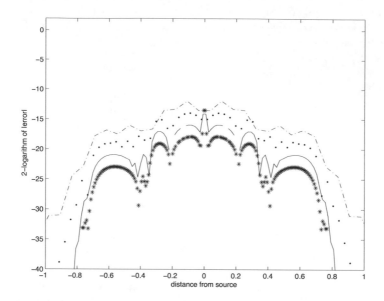

FIGURE 6.35: The 2-logarithm of the error along a line going through the source point for a point force located at $x = 0, y = 0$, and approximated in space by (6.225). Note that the error only decays as $\mathcal{O}(h^2)$ away from the source. The grid sizes were $h = 0.05\,(-\cdot)$, $0.025\,(\cdot)$, $0.0125\,(-)$, $0.00625\,(*)$.

TABLE 6.67:   Errors in max-norm for decreasing h and analytical solution U^{true}. Convergence rate approaches fourth-order after the singular forcing term goes to zero of the two-dimensional split scheme.

Grid step h	Time $t = 1.1$, $e_h = \mathrm{err}_{U,L_\infty} = \lVert U^n - U^{\mathrm{true}} \rVert_\infty$ case 1	$\log_2\!\left(\frac{e_{2h}}{e_h}\right)$
0.05	1.1788e-04	
0.025	1.4146e-05	3.0588
0.0125	1.3554e-06	3.3836
0.00625	1.0718e-07	3.6606
0.003125	7.1890e-09	3.8981

A visualization of this elastic wave propagation is given in Figures 6.36 and 6.37.

To visualize the influence of the singular point force for the 2D computation, we use the x-component of the solution at time $t = 1$ over the normed domain $[-1, 1] \times [-1, 1]$.

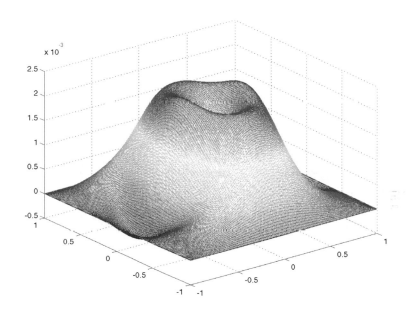

FIGURE 6.36: The x-component of the solution for a singular point force at time $t = 1$ and spatial grid step $h = 0.0125$.

6.5.2.7 Computational Cost of the Splitting Method

For a two-dimensional problem, the fourth-order explicit method (6.210) can be implemented using approximately 160 *floating point operations* (flops) per grid point.

The splitting method requires approximately 120 flops (first step) plus 2-times 68 flops (second and third step) for a total of 256 flops. This increase of about 60% in the number of flops is somewhat offset by the larger time steps allowed by the splitting method, especially for "nice" material properties, making the two methods roughly comparable in computational cost.

6.5.2.8 A Three-Dimensional Splitting Method

As for the splitting method discussed for two dimensions in Section 6.5.2.5, we extend the splitting method to three dimensions in the applications. Therefore, a first discussion of the discretization and decomposition of the operator is important. In three dimensions, a fourth-order difference approximation of the operator becomes

$$
\mathcal{M}_4 = \left(\begin{array}{c}
(\lambda + 2\mu)\left(1 - \tfrac{h^2}{12}D^{x2}\right)D^{x2} + \mu\left(1 - \tfrac{h^2}{12}D^{y2}D^{y2} + 1 - \tfrac{h^2}{12}D^{z2}\right)D^{z2} \\
(\lambda + \mu)\left(1 - \tfrac{h^2}{6}D^{x2}\right)D_0^x\left(1 - \tfrac{h^2}{6}D^{y2}\right)D_0^y \\
(\lambda + \mu)\left(1 - \tfrac{h^2}{6}D^{x2}\right)D_0^x\left(1 - \tfrac{h^2}{6}D^{z2}\right)D_0^z
\end{array} \right.
$$

$$
\begin{array}{c}
(\lambda + \mu)\left(1 - \tfrac{h^2}{6}D^{x2}\right)D_0^x\left(1 - \tfrac{h^2}{6}D^{y2}\right)D_0^y \\
(\lambda + 2\mu)\left(1 - \tfrac{h^2}{12}D^{y2}\right)D^{y2} + \mu\left(1 - \tfrac{h^2}{12}D^{x2}\right)D^{x2} + \mu\left(1 - \tfrac{h^2}{12}D^{z2}\right)D^{z2} \\
(\lambda + \mu)\left(1 - \tfrac{h^2}{6}D^{z2}\right)D_0^z\left(1 - \tfrac{h^2}{6}D^{y2}\right)D_0^y
\end{array}
$$

$$
\left. \begin{array}{c}
(\lambda + \mu)\left(1 - \tfrac{h^2}{6}D^{x2}\right)D_0^x\left(1 - \tfrac{h^2}{6}D^{z2}\right)D_0^z \\
(\lambda + \mu)\left(1 - \tfrac{h^2}{6}D^{y2}\right)D_0^y\left(1 - \tfrac{h^2}{6}D^{z2}\right)D_0^z \\
(\lambda + 2\mu)\left(1 - \tfrac{h^2}{12}D^{z2}D^{z2} + \mu\left(1 - \tfrac{h^2}{12}D^{x2}\right)D^{x2} + 1 - \tfrac{h^2}{12}D^{y2}\right)D^{y2}
\end{array} \right),
$$

$$(6.228)$$

where D^{x2}, D^{y2}, D^{z2} are central difference approximations of the second derivatives $\partial_x^2, \partial_y^2, \partial_z^2$, and D_0^x, D_0^y, D_0^z are central difference approximations of the first derivatives $\partial_x, \partial_y, \partial_z$.

The operator (6.228) is applied to grid functions $U_{j,k,l}^n$ defined at grid points x_j, y_k, z_l, t_n similarly to the two-dimensional case.

We can split $\mathcal{M}_4$ into six parts; $\mathcal{M}_{xx}, \mathcal{M}_{yy}, \mathcal{M}_{zz}$ containing the three second-order directional difference operators, and $\mathcal{M}_{xy}, \mathcal{M}_{yz}, \mathcal{M}_{xz}$ containing the mixed difference operators.

We could split this scheme in a number of different ways depending on how we treat the mixed derivative terms. We have chosen to implement the following split scheme in three dimensions:

1. $$\rho \frac{U^{*}_{j,k,l} - 2U^{n}_{j,k,l} + U^{n-1}_{j,k,l}}{\tau_n^2} = \mathcal{M}_4 U^{n}_{j,k,l} \tag{6.229}$$

$$+\theta f^{n+1}_{j,k,l} + (1 - 2\theta) f^{n}_{j,k,l} + \theta f^{n-1}_{j,k,l}),$$

2. $$\rho \frac{U^{**}_{j,k,l} - U^{*}_{j,k,l}}{\tau_n^2} = \theta \mathcal{M}_{xx} \left(U^{**}_{j,k,l} - 2U^{n}_{j,k,l} + U^{n-1}_{j,k,l} \right) \tag{6.230}$$

$$+\frac{\theta}{2} \left(\mathcal{M}_{xy} + \mathcal{M}_{xz} \right) \left(U^{*}_{j,k,l} - 2U^{n}_{j,k,l} + U^{n-1}_{j,k,l} \right),$$

3. $$\rho \frac{U^{***}_{j,k,l} - U^{**}_{j,k,l}}{\tau_n^2} = \theta \mathcal{M}_{xx} \left(U^{***}_{j,k,l} - 2U^{n}_{j,k,l} + U^{n-1}_{j,k,l} \right) \tag{6.231}$$

$$+\frac{\theta}{2} \left(\mathcal{M}_{xy} + \mathcal{M}_{yz} \right) \left(U^{**}_{j,k,l} - 2U^{n}_{j,k,l} + U^{n-1}_{j,k,l} \right),$$

4. $$\rho \frac{U^{n+1}_{j,k,l} - U^{***}_{j,k,l}}{\tau_n^2} = \theta \mathcal{M}_{xx} \left(U^{n+1}_{j,k,l} - 2U^{n}_{j,k,l} + U^{n-1}_{j,k,l} \right) \tag{6.232}$$

$$+\frac{\theta}{2} \left(\mathcal{M}_{xz} + \mathcal{M}_{yz} \right) \left(U^{***}_{j,k,l} - 2U^{n}_{j,k,l} + U^{n-1}_{j,k,l} \right).$$

Properties such as the splitting error, accuracy, stability, etc., for the three-dimensional case are similar to those of the two-dimensional case treated in the earlier sections.

6.5.2.9 Test Example of the 3D Wave Equation

We have performed some numerical experiments with the three-dimensional scheme in order to test its convergence and stability. We apply our splitting scheme, presented in Subsubsection 6.5.2.8, and use a forcing

$$\begin{aligned} f = \big(&-(-1 + \lambda + 4\mu)\sin(t - x)\sin(y)\sin(z) - \\ &(\lambda + \mu)\cos(x)(2\sin(t)\sin(y)\sin(z) + \cos(t)\sin(y + z)), \\ &-(-1 + \lambda + 4\mu)\sin(x)\sin(t - y)\sin(z) - \\ &(\lambda + \mu)\cos(y)(2\sin(t)\sin(x)\sin(z) + \cos(t)\sin(x + z)), \\ &-(\lambda + \mu)\cos(t - y)\cos(z)\sin(x) - \sin(y)((\lambda + \mu)\cos(t - x)\cos(z) + \\ &(-1 + \lambda + 4\mu)\sin(x)\sin(t - z))\big)^T, \end{aligned} \tag{6.233}$$

giving the analytical solution

$$\begin{aligned} U^{\text{true}} = \big(&\sin(x - t)\sin(y)\sin(z), \\ &\sin(y - t)\sin(x)\sin(z), \\ &\sin(z - t)\sin(x)\sin(y)\big)^T. \end{aligned} \tag{6.234}$$

As in the earlier examples, we tested this for a number of different grid sizes. Using the same two sets of material parameters as for the two-dimensional

TABLE 6.68:   Errors in max-norm for decreasing h
and smooth analytical solution U^{true}. The convergence
rate indicates fourth-order convergence of the
three-dimensional split scheme.

h	$t = 2$, $e_h = \text{err}_{U, L_\infty} = \|U^n - U^{\text{true}}\|_\infty$			
	case 1	$\log_2\left(\frac{e_{2h}}{e_h}\right)$	case 2	$\log_2\left(\frac{e_{2h}}{e_h}\right)$
0.1	4.2986e-07		1.8542e-06	
0.05	3.5215e-08	3.61	1.3605e-07	3.77
0.025	3.0489e-09	3.53	8.0969e-09	4.07
0.0125	2.0428e-10	3.90	4.7053e-10	4.10

case, we computed up to $t = 2$ and checked the maximum error for all components of the solution. The results are given in Table 6.68.

We implemented the 3D case in MATLAB$^{\circledR}$ on a single CPU. Further computations were also done on multiple CPUs with optimal speed.

The spatial grid steps are given as:
$h = \Delta x = \Delta y = \Delta y \in \{0.1, 0.05, 0.025, 0.0125\}$.
Further, the time step is given as $\tau = \tau = 0.0046$. The parameters for this elastic wave propagation are $\lambda = 14$, $\mu = 1$, and $\rho = 1$. We obtain a total number of grid points of about $1\ 10^6$.

The initial solution is given as $U = 0$.

In Figure 6.37, we visualize the influence of the singular point force for the 3D computation using the y-component on a plane. The solution is given at the time $t = 1$ and at the normed domain $[-1, 1] \times [-1, 1] \times [-1, 1]$. Furthermore, the isosurfaces for the solution U are also visualized to show the wave fronts.

Remark 6.31. Our splitting scheme has been shown to work well in practice for different types of material properties. It is comparable to the fully explicit fourth-order scheme (6.210) in terms of computational cost, but should be easier to implement, as no difference approximations of higher-order operators are needed. To improve the accuracy and reduce the time splitting error, we have also applied an embedded iterative splitting scheme, see also [85] and [92]. Here, a further problem is to reduce the spatial discretization error with fine spatial grids or more accurate schemes, before reducing the time discretization error, see [77].

A vital component of a model in seismology is the stable higher-order approximation of the boundary conditions. Such implementation can also be done in a more general setting; some cases are discussed in [85].

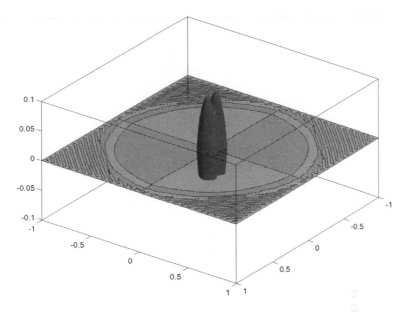

FIGURE 6.37: Contour plot of the y-component on a plane for the 3D case with a singular force. The isosurface for $||U||$ is also included.

6.5.3 CVD Apparatus: Optimization of a Deposition Problem

We base our study on controlling the growth of a thin film that can be done by CVD (chemical vapor deposition) processes [162, 170, 179]. In recent years, the production of thin films by low temperature and pressure deposition has increased enormously. The low temperature and low pressure allow one to deposit a new class of so-called MAX-phase material [12], which has high electrical conductivity and heat conductivity characteristics on cheap and well-known metallic materials, e.g., steel. Such critical processes have to be simulated with knowledge of the underlying transport models to control the positions and forms of the sources, e.g., point, fractal, or line sources.

We apply such models as are related to mesoscopic scale modeling, [115], with respect to flows close to the wafer surface. Such a wafer surface can be modeled as a porous substrate, [185]. A further model is done by considering the connection between the CVD process and porous media, while dealing with porous ceramic membranes and/or gas catalysts (Argon, Xenon), which are porous medias. These membranes and gas catalysts are inlets in the chamber of the CVD apparatus and the gas stream has to go through these porous media, [25]. For membrane applications, proper modeling in pore size (distribution) is essential. We concentrate on modeling the gas chamber with a porous medium inlet and consider the target substrate as a further porous medium. We assume a stationary homogeneous and nonionized plasma that does not interact with the gaseous species, [162]. For such a homogenous plasma we have applied our expertise in modeling transport through a porous medium. The first model is based on the gas mixture velocity that can be simulated by the flow through the chamber (homogeneous porous medium) and the flow into the porous substrate (target). These techniques are the initial stage of the electrochemical vapor deposition (EVD) introduced by Westinghouse Electric Corporation and discussed in [203] for the deposition of a thin dense layer on the top of a porous substrate. To improve the modeling of the gaseous flow to the gas chamber, we deal with the so-called far-field model based on a porous media. Here the plasma can be modeled as a continuous flow [115]. We assume a near vacuum and a diffusion-dominated process, derived from the Knudsen diffusion, [25]. In such viscous flow regimes, we deal with small Knudsen Numbers and a pressure of nearly zero.

In Figure 6.38, the gas chamber of the CVD apparatus is shown, which is done with a porous media.

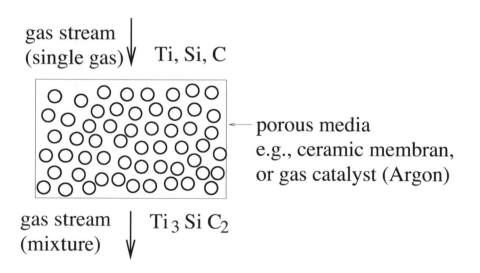

FIGURE 6.38: Gas chamber of the CVD apparatus.

To solve the model, we concentrate on discretization methods that take into account the individual time and spatial scale of each part of the equation. The reaction and source terms can be treated with fast Runge-Kutta (RK) methods, where the convection-diffusion parts are solved with splitting semi-implicit finite volume methods and characteristic methods, [73].

The numerical results treat the applications in the production of so-called metallic bipolar plates. Next we discuss mathematical model.

6.5.3.1 Mathematical Model

In the following, the models are discussed in terms of far-field and near-field problems:

1. Reaction-diffusion equations [115] (far-field problem);

2. Boltzmann-lattice equations [190] (near-field problem).

The modeling is characterized by a Knudsen Number, which is the ratio of the mean free path λ to the typical domain size L. For small Knudsen Numbers $Kn \approx 0.01 - 1.0$ we deal with a Navier-Stokes equation, whereas for large Knudsen numbers $Kn \geq 1.0$ we deal with a Boltzmann equation [179].

We concentrate on a far-field model and assume a continuum flow, and that the fluid equations can be treated with a Navier-Stokes equation. For simplification we deal with a constant velocity field. The transport of the species in the gas chamber is modeled with a convection-diffusion-reaction equation, see:

$$\partial_t u + \nabla \cdot F - R_g = 0, \text{ in } \Omega \times [0, t], \tag{6.235}$$
$$F = \mathbf{v}u - D\nabla u,$$
$$u(x, t) = u_0(x), \text{ on } \Omega, \tag{6.236}$$
$$u(x, t) = u_1(x, t), \text{ on } \partial\Omega \times [0, t], \tag{6.237}$$

where u is the molar concentration and F is the flux of the species. $\mathbf{v}$ is the flux velocity through the chamber and porous substrate [185]. D is the diffusion matrix and R_g is the reaction term. The initial value is given as u_0 and we assume a Dirichlet boundary condition with the function $u_1(x, t)$ sufficiently smooth.

The diffusion in the modified CVD process is given by the Knudsen diffusion, [25]. We consider the overall pressure in the reactor to be 200 Pa and the substrate temperature to be about $600 - 900$ K. The pore size in the porous substrate is 80 nm (nanometer).

The diffusion is described by:

$$D = \frac{2\epsilon\mu_K \mathbf{v}r}{3RT}, \tag{6.238}$$

where ϵ is the porosity, μ_K is the shape factor of the Knudsen diffusion, r is the average pore radius, R and T are the gas constant and temperature, respectively, and $\mathbf{v}$ is the mean molecular speed, given by:

$$\mathbf{v} = \sqrt{\frac{8RT}{\pi W}}, \tag{6.239}$$

where W is the molar mass of the diffusive gas.

For the homogeneous reactions, during the CVD process we take a constant reaction of Si, Ti, and C given by:

$$3Si + Ti + 2C \rightarrow Si_3TiC_2, \tag{6.240}$$

where Si_3TiC_2 is the MAX-phase material that deposits at the target.

For simplicity, we do not consider the intermediate reaction with the precursor gases, [162].

The reaction rate is then given by:

$$\lambda = k_r \frac{[3Si]^M [Ti]^N [2C]^O}{[Si_3TiC_2]^L}, \qquad (6.241)$$

where k_r is the apparent reaction constant, and L, M, N, O are the reaction orders of the reactants.

The velocity in the homogeneous substrate is modeled by a porous medium [17, 141]. We have assumed a stationary medium, e.g., nonionized plasma or nonreactive precursor gas. Furthermore, the pressure can be assumed with the Maxwell distribution as [162]:

$$p = \rho b T, \qquad (6.242)$$

where b is Boltzmann constant and T is the temperature.

We have modeled the velocity by partial differential equations. Here we assume the gaseous flow is a nearly liquid flow through the porous medium. We can therefore derive by Darcy's law:

$$\mathbf{v} = -\frac{k}{\mu}(\nabla p - \rho g), \qquad (6.243)$$

where $\mathbf{v}$ is the velocity of the fluid, k is the permeability tensor, μ is the dynamic viscosity, p is the pressure, g is the vector of the gravity, and ρ is the density of the fluid.

We use the continuum equation of the particle density and obtain the equation of the system, which is taken as our flow equation:

$$\partial_t(\phi\rho) + \nabla \cdot (\rho\mathbf{v}) = Q, \qquad (6.244)$$

where ρ is the unknown particle density, ϕ is the effective porosity, and Q is the source term of the fluid. We assume a stationary fluid and consider only divergence-free velocity fields, i.e.,

$$\nabla \cdot \mathbf{v}(x) = 0, \quad x \in \Omega. \qquad (6.245)$$

The boundary conditions for the flow equation are given by:

$$p \;=\; p_r(t, \gamma), \quad t > 0, \quad \gamma \in \partial\Omega, \qquad (6.246)$$
$$\mathbf{n} \cdot \mathbf{v} \;=\; m_f(t, \gamma), \quad t > 0, \quad \gamma \in \partial\Omega, \qquad (6.247)$$

where $\mathbf{n}$ is the normal unit vector with respect to $\partial\Omega$ and we assume that the pressure p_r and flow concentration m_f are prescribed by Dirichlet boundary conditions [141].

From the stationary fluid, we assume that conservation of momentum for velocity $\mathbf{v}$ holds [114, 141]. So we can neglect the computation of the momentum for the velocity.

Remark 6.32. For the flow through the gas chamber, for which we assume a homogeneous medium and nonreactive plasma, we have considered a constant flow [129]. A further simplification is given by the very small porous substrate, for which we can assume the underlying velocity in a first approximation as being constant [179].

Remark 6.33. For a nonstationary medium and reactive or ionized plasma, we have to take into account the relations of the electrons in the thermal equilibrium. Such spatial variation can be considered by modeling the electron drift. Such modeling of the ionized plasma is done with Boltzmann relation, [162].

6.5.3.2 Parameters of the Two-Dimensional Simulations

In the following we assume the transport and reaction of a single species, where A is the silicon species and B is the end product Ti_3SiC_2 (Max-phase species). The parameters of the evolution equation, see Section 1.2.3, are given in Table 6.69.

TABLE 6.69: Model parameters for the two-dimensional simulations.

Density	$\rho = 1.0$
Mobile porosity	$\phi = 0.333$
Immobile porosity	$\phi_{im} = 0.0$
Diffusion	$D = 0.0$
Longitudinal dispersion	$\alpha_L = 0.0$
Transversal dispersion	$\alpha_T = 5.0$
Retardation factor	$R = 10.0 \ 10^{-4}$ (Henry rate)
Velocity field	$\mathbf{v} = (0.0, -4.0 \ 10^{-8})^t$
Decay rate of species of first species	$\lambda_{A \to B} = 2 \ 10^{-8}$
Geometry (2D domain)	$\Omega = [0, 100] \times [0, 100]$
Boundary	Neumann boundary on top, left, and right boundaries Outflow boundary on the bottom boundary

For the discretization methods of the coupled transport-reaction equation, we use, for the spatial discretization, a finite volume method for unstructured grids. For the time discretization, we apply Crank-Nicolson or fractional step methods.

The finite volume method of the second order has the following parameters, described in Table 6.70.

The time discretization methods of the second order (Crank-Nicolson) has the following parameters, described in Table 6.71.

For the solution method, which is a combination of CG (conjugate gradient) and the multigrid method, see [8] and [119], we have also embedded a

TABLE 6.70:   Spatial discretization parameters for the two-dimensional simulations.

Spatial step size	$\Delta x_{min} = 1.56, \Delta x_{max} = 2.21$
Refined levels	6
Limiter	Slope limiter
Test functions	Linear test function
	Reconstructed with neighbor gradients

TABLE 6.71:   Time discretization parameters for the two-dimensional simulations.

Initial time step	$\Delta t_{init} = 5\ 10^7$
Controlled time step	$\Delta t_{max} = 1.298\ 10^7, \Delta t_{min} = 1.158\ 10^7$
Number of time steps	$100, 80, 30, 25$
Time-step control	Time steps are controlled with
	the Courant number $\text{CFL}_{max} = 1$

splitting method to decouple the problem into a transport part and a reaction part. Here the transport part is solved after spatial discretization with iterative solution schemes, while the reaction part is an ordinary differential equation and can be solved analytically or with fast ODE solvers.

The solver parameters are given in Table 6.72.

TABLE 6.72:   Solution methods and their parameters for the two-dimensional simulations.

Solver	BiCGstab (biconjugate gradient method)
Preconditioner	Geometric multigrid method
Smoother	Gauss-Seidel method as smoother for
	multigrid method
Basic level	0
Initial grid	Uniform grid with two elements
Maximum level	six
Finest grid	Uniform grid with 8192 elements

Remark 6.34. We also applied a more refined grid and obtained at least the same needed accuracy. The splitting method is embedded and we solve the reaction part analytically, whereas the transport part is solved numerically. For the iterative scheme we choose two iterative steps, because of the underlying second-order time and spatial discretization scheme.

Such an order barrier can be improved by higher-order time and spatial discretization schemes, such that it makes sense to improve additional the iterative steps, see the discussion in [82].

In the following section we describe the optimization of a deposition method.

6.5.3.3   Experiments with the Plasma Reactor (Two-Dimensional)

In the following we discuss the projects of the plasma reactor.

We discuss the optimization of layer growth and the optimal growth rate of a single species, and we propose in our simulation that C carbon dominates the process for the Max-phase materials [12].

Optimization of layer growth

In this project, we optimize the underlying layer growth with respect to the sources of the deposition material. We have just one source to deposit on the target and to simulate the dominant species, i.e., carbon. Our intention is to get a homogeneous layer at the target. An industrial partner is needed to replace gold or silver, which can be controlled to provide a homogeneous layer at the target [189]. Financial constraints impose the requirement of a cheap material, e.g., Max-phase material [12].

In our contribution we present ideas for obtaining the homogeneous layer.

In Figure 6.39, we show the optimal deposition rates to obtain a homogeneous layer.

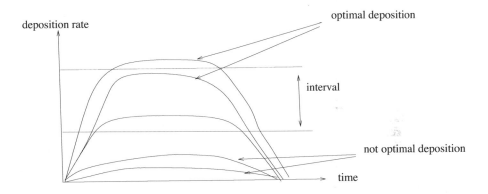

FIGURE 6.39: Amount of optimal deposition rates with respect to the deposition time.

We contribute the following ideas:

Idea 1

We propose to optimize the source of the deposition material with respect to a time-dependent source that moves up and down in the y-direction. So we obtain a deposition along the length of the target layer.

Here we simulate two experiments, in each of which the time steps are equal to 200.

In the first experiment, we take the moving distance steps to be equal to five, and in the second experiment we take the moving distance step to be equal to ten.

In Figure 6.40, we present the experiment of a single source with moving distance steps equal to five.

In Figure 6.41, we present the deposition rates of the experiment of a single source with moving distance steps equal to five.

In Figure 6.42, we present the deposition rates of the second experiment of the single source with larger moving steps.

Remark 6.35. The two experiments yield different deposition rates. Smaller moving steps yield an improved deposition, so we can see the more homogeneous rates in the first experiment. The comparison between Figures 6.41 and 6.42 shows that Figure 6.41 gives us a better result in the deposition rates than Figure 6.42. It is nearly stable and has no peaks.

In the future we will test other moving distance steps to obtain the best result.

Idea 2

We propose to take many small sources, so called point sources, in the y-direction.

Here, we obtain over the whole target layer the influence of the deposition material, but for the whole time period.

We performed many experiments, in all of which the time steps are equal to $\Delta t = 200$. The concentration of each source alternates from 1 to 0 and from 0 to 1 during a small time interval. In the following we will see the results of just two experiments, the parameters of the source positions are given in Table 6.73.

TABLE 6.73:   Experiments with the second deposition idea for the two-dimensional simulations.

No. of Experiment	No. of Sources	Source Positions	Distance of Sources
1	3	$(50.50), (50.40), (50.60)$	10
2	5	$(50.40)(50.45)(50.50)(50,55)(50.60)$	5

In Figure 6.43, we present the deposition rate of the first experiment with the second idea when it involves three single sources.

In Figure 6.44, we present the deposition rate produced by the second idea when it involves five point sources and different positions of the sources.

Remark 6.36. The amount of sources plays an important role in enlarging the deposition time and improving the layering. Here, with five sources, we obtain a more homogeneous profile of the deposition. The comparison between Figures 6.43 and 6.44 shows that both figures have nearly the same results. The deposition rates are not stable and we obtain many peaks. Incidentally, Figure 6.44 has higher deposition rates that are embedded in the optimal rates (see Figure 6.39).

In the future we will use a greater number of point sources to obtain the best homogeneous layer at the target.

Idea 3

In the next experiments we propose an area source in the x-y plane.

The variations occur in the areas of the sources and the different distances from the target layer.

We did two experiments with far-field sources and time steps equal to 200, and two experiments with near-field sources and time steps equal to 250. The parameters of the area sources are given in Table 6.74.

TABLE 6.74: Experiments with the third idea of an area source for the two-dimensional simulations.

No. of Experiment	Type of Sources	Area of Sources	Time steps Δt
1	Fare	$(50, 55) \times (30, 70)$	200
2	Fare	$(50, 55) \times (40, 60)$	200
3	Near	$(75, 80) \times (20, 80)$	250
4	Near	$(75, 80) \times (40, 60)$	250

In the first experiment, we take a value of x between 50 and 55, and a value of y between 30 and 70.

In Figure 6.45, we present the large area source when the value of x ranges from 50 to 55, and the value of y ranges from 30 to 70 with time steps equal to 200.

In Figure 6.46, we present the deposition rate of the large area source when the value of x ranges from 50 to 55, and the value of y ranges from 30 to 70 with time steps equal to 200.

In the next experiments, we take a value of x from 75 to 80 and a value of y from 20 to 80.

In Figure 6.47, we present the deposition rate of a large area source.

In the last experiment, we take a value of x from 75 to 80, and a value of y from 40 to 60.

In Figure 6.48, we present a small area source when the value of x is from 75 to 80, and the value of y is from 40 to 60 with time steps equal to 250.

In Figure 6.49, we present the deposition rate of the small area source.

Remark 6.37. In the last experiment, we obtained the benefit of an area source that can be controlled to deposit a homogeneous layer of the new material. The controlling of the deposition rate can be done more simply with area sources that cover the deposition area, than with moving sources or multiple point sources. The deposition rates are covered in the constraint interval.

Incidentally, we may compare Figures 6.41, 6.44, and 6.47 and we see that the best result is from Figure 6.47, which is very stable and has a homogeneous layer.

time (in ns) : 2.472E+07

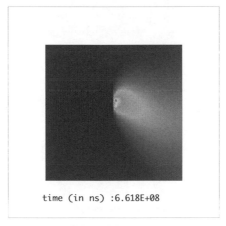

time (in ns) :6.618E+08

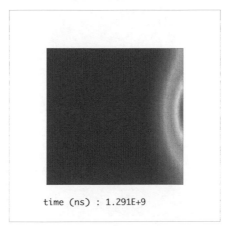

time (ns) : 1.291E+9

FIGURE 6.40: Source is moving in the y-direction with moving distance step $\Delta y = 5$.

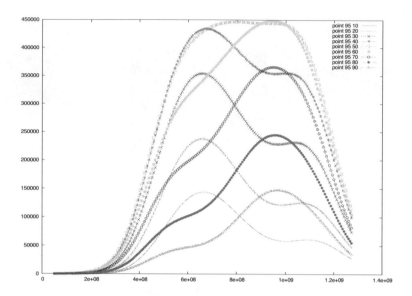

FIGURE 6.41: Deposition rates for the case of a source moving in the y-direction with moving distance step 5.

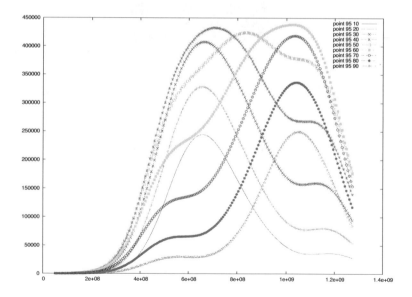

FIGURE 6.42: Deposition rates for the case of a source moving in the y-direction with step $\delta y = 10$.

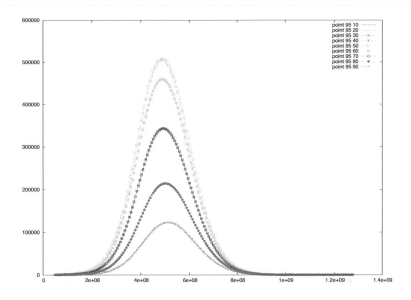

FIGURE 6.43: Deposition rates in the case of three point sources in the y-direction with moving distance step $\delta y = 10$ and $x = 50$.

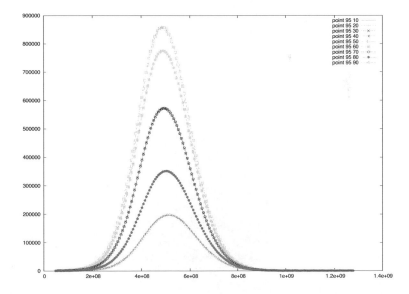

FIGURE 6.44: Deposition rates in the case of five point sources in the y-direction with moving distance step $\delta y = 10$, $x = 50$.

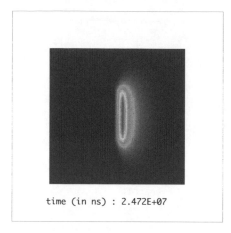

time (in ns) : 2.472E+07

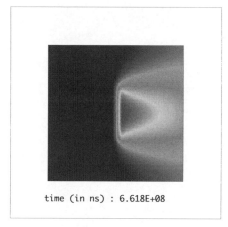

time (in ns) : 6.618E+08

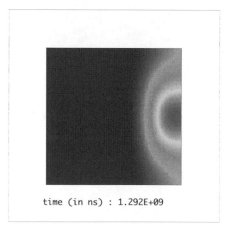

time (in ns) : 1.292E+09

FIGURE 6.45: Line source, x between 50 and 55, y between 30 and 70.

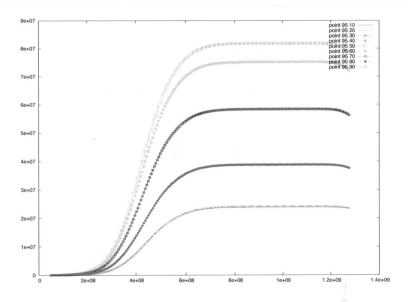

FIGURE 6.46: Deposition rates in the case of a line source, x between 50 and 55, y between 30 and 70.

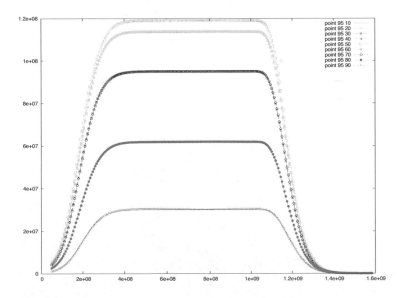

FIGURE 6.47: Deposition rates in the case of a line source, x between 75 and 80, y between 20 and 80.

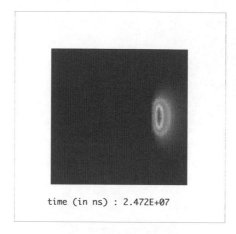

time (in ns) : 2.472E+07

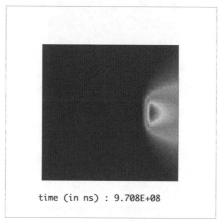

time (in ns) : 9.708E+08

time (in ns) : 1.601E+09

FIGURE 6.48: Line source, x between 75 and 80, y between 40 and 60.

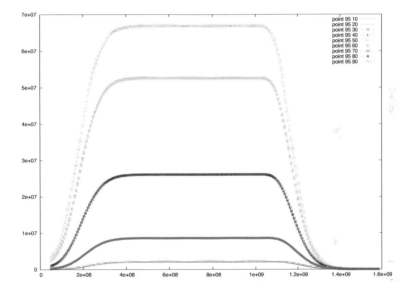

FIGURE 6.49: Deposition rates in the case of a line source, x between 75 and 80 y, between 40 and 60.

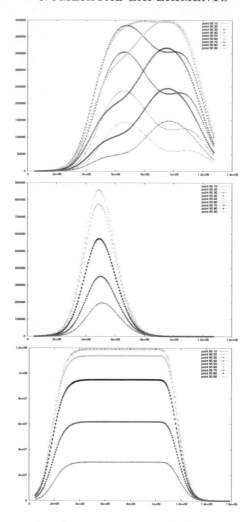

FIGURE 6.50: Comparison of the various sources.

Remark 6.38. Figure 6.50 compares the deposition rates resulting from the three ideas: we can see that the best result is achieved with the third idea –a line source– but it is more expensive than the other ideas; in the future we will try a combination of the point sources and the line sources, or a combination of all three ideas. That means we can take many point sources and let these sources move in time.

Conclusions

We present a continuous and kinetic model involving the far-field effect.

The connection between transport through a gas chamber and a porous substrate is shown by the use of plasma as a gaseous substrate. We explain the different scale models and can predict the flow of the reacting chemicals on the scale of the chemical reactor. For the mesoscopic scale model we demonstrate the multiscale discretization and solver methods. Numerical examples are presented to discuss the influence of the near-continuum regime at the thin film. The modeling of various inflow sources can describe the growth of the thin film at the wafer. As a result, a combination of multiple point and line sources can control the delicate growth process and yield a homogeneous layer at the target.

6.5.3.4   Parameters of the Three-Dimensional Simulations

In the following we assume the transport and reaction of a single species, where A is the silicon species and B is the end product Ti_3SiC_2 (Max-phase species).

For computational reasons, we scale all the parameters to the velocity $(v_x, v_y, v_z)^t = (0, -1, 0)^t$. For our main goal, of verification, we need only qualitative results, but we can also shift the results to the two-dimensional parameters.

The parameters of the transport-reaction equations, discussed in Section 1.2.3, are given in Table 6.75.

TABLE 6.75:   Model parameters for the three-dimensional simulations.

Density	$\rho = 1.0$
Mobile porosity	$\phi = 0.333$
Immobile porosity	0.0
Diffusion	$D = 1.0 \ 10^{-3}$
Longitudinal dispersion	$\alpha_L = 1.0$
Transversal dispersion	$\alpha_T = 1.0$
Retardation factor	$R = 1.0$ (Henry rate)
Velocity field	$\mathbf{v} = (0.0, -1.0, 0.0)^t$
Decay rate of species of first species	$\lambda_{A \to B} = 2 \ 10^{-8}$
Geometry (3D domain)	$\Omega = [0, 100] \times [0, 100] \times [0, 100]$
Boundary	Neumann boundary on top, left, and right boundaries Outflow boundary on the bottom boundary

We apply finite volume methods to achieve the spatial discretization and the Crank-Nicolson method for the time discretization.

The parameters of the finite volume method of the second order are given in Table 6.76.

TABLE 6.76: Spatial discretization parameters for the three-dimensional simulations.

Spatial step size	$\Delta x_{min} = 1.56, \Delta x_{max} = 43.3$
Refined levels	6
Limiter	Slope limiter
Test functions	Linear test function
	Reconstructed with neighbor gradients

For the time discretization method, we apply the Crank-Nicolson method (second order): the parameters are given in Table 6.77.

TABLE 6.77: Time discretization parameters for the three-dimensional simulations.

Initial time step	$\Delta t_{init} = 0.1$
Controlled time step	$\Delta t_{max} = 0.1, \Delta t_{min} = 1.0$
Number of time steps	200
Time-step control	Time steps are controlled with
	The Courant-number $\mathrm{CFL}_{max} = 1$

For the solution method, it is important to decouple the problem into a PDE part and an ODE part. While the PDE part (the transport equation) is solved after spatial discretization with time-consuming iterative solvers, the ODE part (the reaction equation) is solved analytically, and therefore cheaply.

The solver parameters are given in Table 6.78.

TABLE 6.78: Solver methods and their parameters for the three-dimensional simulations.

Solver	BiCGstab (Biconjugate gradient method)
Preconditioner	Geometric multigrid method
Smoother	Gauss-Seidel method as smoother for
	Multigrid method
Basic level	0
Initial grid	Uniform grid with 40 elements
Maximum level	six
Finest grid	Adaptively refined grid with 164034 elements

Remark 6.39. We applied an error indicator for convection and diffusion equations based on flux differences. We use the method of local mesh refinement, which is controlled by the numerical solution itself on the basis of *a posteriori* error estimates. They can measure the difference between the exact and the numerical solution in terms of the numerical solution that is

known, see [62]. A grid element is refined while the difference between the numerical solution between the neighboring elements is large in comparison to the defined error bound [178].

Remark 6.40. We consider as in the two-dimensional case, that the splitting method is embedded and we solve the reaction part analytically or semi-analytically, whereas the transport part is solved numerically. For the iterative scheme we choose maximal two iterative steps, because of the underlying second-order time and spatial discretization scheme.

6.5.3.5 Experiments with the Plasma Reactor (Three-Dimensional)

In the following we extend the discussion to the simulation of a plasma reactor in three dimensions.

We assume that because of symmetry and the lesser influence of anisotropy effects, we can save on computational time and reduce all experiments to two dimensions.

Here we have the following benefits, which can be extended to the influence of the transport and reaction process in the gas phase:

- Verification of the two-dimensional results (e.g., Are the assumptions of the models correct?)

- Additional effects in the third dimension (Should we consider any anisotropy that would make the third dimension important?)

- Additional tests of all two-dimensional problems, that the methods should work well also in three-dimensional problems (Can we obtain similar three-dimensional effects?)

In the next experiments, we compare deposition of the species with the underlying deposition rates. Such experiments help us to see homogeneous depositions, see Figure 6.39.

Experiment (Source Points)

The underlying experiment deals with point sources and computes the deposition rates that result. We add additional point sources to improve the homogeneity of the deposition process.

We concentrate on a single species that reacts to the target material. Furthermore, we assume we have a homogeneous medium and that the gas flux is constant in the small deposition area.

With these assumptions, the interest is only focused on the diffusive flux and the consequences for a homogeneous deposition with point sources.

While in two-dimensional experiments, we need a huge number of point sources to obtain a good deposition result. Nevertheless, here we consider a simplified experiment with three point sources.

In Figure 6.51, we present the deposition rate of three point sources.

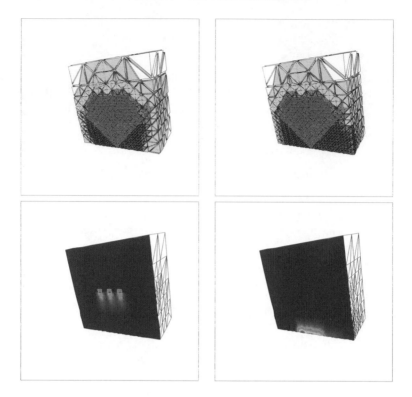

FIGURE 6.51: Point sources: $(40, 50, 50)$, $(50, 50, 50)$, and $(60, 50, 50)$ (left figures: start configuration after 10 time steps, right figures: end configuration after 200 time steps).

Remark 6.41. In our three-dimensional experiments, we can verify the results that we obtained in the two-dimensional experiments. The point sources have the same problems as in the two-dimensional experiments. One could only control in a small area (diffusion cone) a homogeneous layer of the deposition. The controlling of the deposition rate can be extended, while using moving point sources, as shown in the two-dimensional experiments, see Section 6.5.3.3.

The decomposition rates are given in Figure 6.52.

Remark 6.42. Figure 6.52 shows us the qualitative comparability of the deposition rates in the previous two-dimensional simulations: we can see that it is more delicate to obtain a homogeneous layer over the whole deposition area. Here it is necessary to combine a point source with a line source. Experiments in three dimensions are more expensive to compute; we have nearly five times

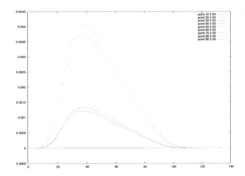

FIGURE 6.52: Decomposition rates in the case of the point sources $(40, 50, 50)$, $(50, 50, 50)$, and $(60, 50, 50)$.

more to do to compute the results.

Experiment (Area Source)

In this experiment, we deal with an area source and compute the deposition rates to a layer that is located at the bottom of the computational domain.

We concentrate on a single species that reacts to the target material. Furthermore, we assume we have a homogeneous medium and that the gas flux is constant in the small deposition area.

With these assumptions, the interest is only focused on the diffusive flux and the consequences for a homogeneous deposition.

Although in the two-dimensional experiments we obtained the best results with line sources, we consider an area source for the three-dimensional case.

To be sharper, the main deposition should be nearly the same area as the given area of the source. So that a loss is only given by the diffusion of the transport process which can be minimized by lower temperatures, see [162].

In Figure 6.53, we present the deposition rate of the small area source.

Remark 6.43. In the three-dimensional experiments, we verify our results that we obtain in the two-dimensional experiments. The area source has the same advantages as the line source in the two-dimensional experiments: one has more control over the deposition of a homogeneous layer of the new material. The controlling of the deposition rate can be done more simply with area sources, because the influence of the diffusion is more negligible than with simple point sources.

The decomposition rates are given in Figure 6.54.

Remark 6.44. Figure 6.54 compares the deposition rates in the previous three ideas of the two-dimensional simulations: We can see that the best

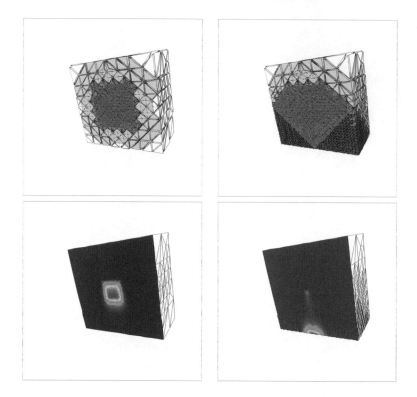

FIGURE 6.53: Area source, x between 40.0 and 60.0, y between 40.0 and 60.0, and z between 40.0 and 60.0 (left figures: start configuration after 10 time steps, right figures: end configuration after 200 time steps).

result is achieved with the third idea –a line source (when two dimensions) or area source (when three dimensions)– but it is more expensive than the other ideas. In the future it will be sufficient to combine point sources and line sources. Experiments in three dimensions are more expensive to compute, we have nearly five times more to do to compute the results.

We perform two- and three-dimensional experiments with at least 200 time steps.

Conclusions

We additionally work with three-dimensional experiments, in order to verify the results of the two-dimensional experiments. For an isotropic diffusion process, the results of the two and three dimensional computations are the same. The splitting method can also applied to three-dimensional problems to decouple the transport and reaction part. Our main result is that the use of line or area sources can produce homogenous layers. The comparison

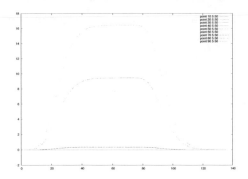

FIGURE 6.54: Decomposition rates in the case of the area source x between $40.0, 60.0$, y between $40.0, 60.0$, and z between $40.0, 60.0$.

TABLE 6.79: Amount of computational time for 2D and 3D experiments at one processor (2,26 GHz, Intel Core 2 Duo processor).

Experiment	Number of elements	Time for one time step
2D (uniform)	131072	0.45 sec.
3D (adaptive)	164016	7.05 sec.

with physical experiments show that multipoint or area sources reduce the inhomogeneity in the growth of thin layers.

6.5.4   Complex Flow Phenomena: Navier-Stokes and Molecular Dynamics

Here the underlying ideas are to couple continuous and discrete problems, which are related to the Navier-Stokes equations and molecular dynamics equations. The methods and experiments are presented in a more detailed way in the preprint Geiser/Steijl [94].

We are motivated by the task of solving for time-dependent fluid flows with complex flow phenomena. In such a computational domain, the Navier-Stokes equations become invalid, e.g., a more detailed computation at the molecular level is needed.

This means that locally, the flow needs to be modeled with a more detailed physical model.

Here the coupling of the Navier-Stokes equations to molecular-level modeling is based on the methods of molecular dynamics.

The Newtonian shear stresses in the Navier-Stokes equations are replaced with the results of molecular dynamics simulations.

The method of molecular dynamics is described in detail in [3], [19], and [197].

This approach is sketched in Figure 1.5, where the Navier-Stokes equations are discretized on a structured cell-centered finite-volume mesh and the solid blocks in the cell face centers close to the upper and lower boundaries denote the microscale problems that are used to evaluate the shear stresses using the molecular dynamics. The coupling of the molecular-level modeling with the discretized continuum system was discussed in [193] for steady flows.

The focus of the present work is on time-dependent flows. Hence, the disparity of the underlying timescales of the continuum flow and molecular-level microscale problems forms the main challenge. In recent years, the topic of multiscale problems has become very large. Solution methods to couple large scales have become important, see [156] for lattice Boltzmann models.

6.5.4.1   Mathematical Model

Our model equations come from fluid dynamics. The macroscale equation is given by the Navier-Stokes equation for an incompressible continuum flow:

$$\rho \partial_t u + \rho(u \cdot \nabla)u - \mu \Delta u + \nabla p = f, \text{in } \Omega \times (0, T), \qquad (6.248)$$

$$\nabla \cdot u = 0, \text{in } \Omega \times (0, T), \qquad (6.249)$$

$$u(0) = u_0, \text{on } \Omega,$$

$$u = 0, \text{on } \partial\Omega \times (0, T).$$

The unknown three-dimensional flow vector
$u = u(x, t) = (u_1(x, t), u_2(x, t), u_3(x, t))^t$ is considered in $\Omega \times (0, T)$ with $x = (x_1, x_2, x_3)^t$. In the above equations, ρ and p are positive constants and represent the fluid density and pressure, respectively. Here, $\mu \in \mathbb{R}^+$ is a constant and represents the dynamic viscosity of the fluid. In the momentum

equation, i.e., Equation (6.248), the term f on the right-hand side represents a volume source term. Equation (6.249) imposes the divergence-free constraint on the velocity field, which is consistent with the assumption of incompressible flow.

The microscopic equation is given by Newton's equation of motion for each individual molecule i for a sample of N molecules,

$$m_i \partial_{tt} x_i = F_i, i = 1, \ldots, N, \tag{6.250}$$

here, x_i is the position vector of atom i, and the force F_i acting on each molecule is the result of the intermolecular interaction of a molecule i with the neighboring molecules within a finite interaction range. In the present work, we assume that the interparticle forces are based on the well-known Lennard-Jones interaction potential [159], i.e., we are assuming the microscopic flow is that of a Lennard-Jones fluid, details of which are given in a later section.

The coupling between the macroscale equation (6.248) and microscale equation (6.250) is assumed to take place through the exchange of the viscous stresses in the momentum equation (6.248). The underlying idea is to replace the viscous stresses based on the Newtonian continuum in Equation (6.248) with a viscous stress evaluated by molecular dynamics simulations of the microscale fluid with the velocity gradient at macroscale level imposed on the microscale fluid, through the use of Lees-Edwards boundary conditions [157]. The molecular-level viscosity is evaluated using the Irving-Kirkwood relation [135].

The viscous stress contribution $\mu \Delta u$ in Equation (6.248) can be generalized for a non-Newtonian flow as $\partial \sigma_{ij}/\partial x_j$, using Einstein summation. In the present work, this non-Newtonian viscous stress contribution is reformulated in the following form:

$$\partial \sigma_{ij}/\partial x_j = \mu_{apparent} \partial^2 u_i/\partial x_j^2, \text{ with } i = 1, 2, 3, \tag{6.251}$$

where, the "apparent" viscosity can be a general function of the imposed velocity gradients in each spatial direction, i.e., this expression can represent general noncontinuum and non-Newtonian flow conditions. The viscous stresses in Equation (6.248) can now be replaced by molecular level viscous stresses by introducing a constant approximate viscosity μ_{approx} and taking into account the deviation of the molecular-level viscous stresses from this approximate viscosity through a volumetric source term.

So, finally, we obtain the coupled multiscale equations:

$$\rho \partial_t u + \rho(u \cdot \nabla)u - \mu_{approx} \Delta u + \nabla p = f, \text{in} \Omega \times (0, T), \tag{6.252}$$
$$f_i = \partial \sigma_{ij}/\partial x_j|_{molecular} - \mu_{approx} \Delta u_i, \text{ with } i = 1, 2, 3,$$

where the Einstein summation convention is used for the volumetric source term f, which accounts for the deviation of the viscous stresses evaluated at the molecular-level from the approximate Newtonian relation $\mu_{approx} \Delta v$. The molecular level viscous stresses can be further reformulated using the "apparent" viscosity, as demonstrated in Equation (6.251).

6.5.4.2 Implicit Dual-Time Stepping Method for Time-Dependent Flows

This section introduces the spatial discretization of the equations governing the flow at the macroscopic scale and at the molecular-level, as well as the coupling method used here. The implicit dual-time stepping method, which forms the basis for the present work, is described in this section. In the next sections, the iterative coupled schemes based on this dual-time stepping method are introduced.

6.5.4.3 Spatial Discretization of Equations for Micro- and Macroscales

The Navier-Stokes equations written in integral form for a domain fixed in time read

$$\frac{d}{dt}\int_{V(t)}\mathbf{w}dV + \int_{\partial V(t)}\left(\mathbf{F}(\mathbf{w}) - \mathbf{F}_v(\mathbf{w})\right)\mathbf{n}dS \;\; = \;\; \mathbf{S}. \qquad (6.253)$$

The above forms a system of conservation laws for any fixed control volume V with boundary ∂V and outward unit normal $\mathbf{n}$. The vector of conserved variables is denoted by $\mathbf{w} = [\rho, \rho u, \rho v, \rho w, \rho E]^T$, where ρ is the density, u, v, w are the Cartesian velocity components, and E is the total internal energy per unit mass. $\mathbf{F}$ and $\mathbf{F}_v$ are the inviscid and viscous flux, respectively. In the absence of volume forces and in an inertial frame of reference the source term $\mathbf{S} = 0$.

Assuming that the x-direction is the homogeneous direction, the y-direction is the cross-flow direction, as well as, that the flow is incompressible with vanishing velocity components in the y- and z-directions, the governing equation can then be written as

$$\rho \partial u/\partial t - \mu_{approx}\partial^2 u/\partial z^2 + \partial p/\partial x = f, \qquad (6.254)$$
$$f = \partial \sigma_{xy}/\partial x - \mu_{approx}\partial^2 u/\partial z^2,$$

using the multiscale equation (6.252).

Equations (6.253) are discretized using a cell-centered finite volume approach on structured multiblock grids, which leads to a set of ordinary differential equations of the form

$$\frac{\partial}{\partial t}\left(\mathbf{w}_{i,j,k}V_{i,j,k}\right) \;\; = \;\; -\mathbf{R}_{i,j,k}\left(\mathbf{w}_{i,j,k}\right), \qquad (6.255)$$

where $\mathbf{w}$ and $\mathbf{R}$ are the vectors of cell variables and residuals, respectively. Here, i, j, k are the cell indices in each block and $V_{i,j,k}$ is the cell volume.

In the present work, the flow problems considered lead to a reduction of the full Navier-Stokes equations to a set of equations for the velocity field components. Hence, the vector $\mathbf{w}$ in this work only involves the Cartesian velocity components, while $\mathbf{R}$ represents the residuals of the three moment equations.

Boundary conditions are imposed by using two layers of halo cells around each grid subdomain. Zero-slip conditions at solid walls are imposed by extrapolating halo cell values in such a way that the velocity at the wall vanishes.

6.5.4.4 Dual-Time Stepping Method

For time-accurate simulations, temporal integration is performed using an implicit dual-time stepping method. Following the pseudotime formulation [138], the updated mean flow solution is calculated by solving the steady state problems

$$
\mathbf{R}_{i,j,k}^* = V_{i,j,k} \frac{3\mathbf{w}_{i,j,k}^{n+1} - 4\mathbf{w}_{i,j,k}^n + \mathbf{w}_{i,j,k}^{n-1}}{2\Delta t} + \mathbf{R}_{i,j,k}\big(\mathbf{w}_{i,j,k}^{n+1}\big) = 0. \quad (6.256)
$$

Equation (6.256) represents a nonlinear system of equations for the full set of Navier-Stokes equations. However, for the reduced system of equations resulting from the homogeneity and periodicity assumptions, presented in Equation (6.254), the system of equations is actually linear. This system is solved by introducing an iteration through *pseudotime* τ to the steady state, as given by

$$
\frac{\mathbf{w}_{i,j,k}^{n+1,m+1} - \mathbf{w}_{i,j,k}^{n+1,m}}{\Delta \tau} + \frac{3\mathbf{w}_{i,j,k}^{n+1,m} - 4\mathbf{w}_{i,j,k}^n + \mathbf{w}_{i,j,k}^{n-1}}{2\Delta t} + \frac{\mathbf{R}_{i,j,k}\big(\mathbf{w}_{i,j,k}^{n+1,m}\big)}{V_{i,j,k}^{n+1}} = 0,
$$

$$(6.257)$$

where the m-th pseudotime iterate at real time step $n + 1$ is denoted by $\mathbf{w}^{n+1,m}$ and the cell volumes are constant during the pseudotime iteration. The unknown $\mathbf{w}_{i,j,k}^{n+1}$ is obtained when the first term in Equation (6.256) converges to a specified tolerance. An implicit scheme is used for the pseudotime integration. The flux residual $\mathbf{R}_{i,j,k}\big(\mathbf{w}_{i,j,k}^{n+1}\big)$ is linearized as follows:

$$
\begin{aligned}
\mathbf{R}_{i,j,k}\big(\mathbf{w}^{n+1}\big) &= \mathbf{R}_{i,j,k}\big(\mathbf{w}_{i,j,k}^n\big) + \frac{\partial \mathbf{R}_{i,j,k}\big(\mathbf{w}_{i,j,k}^n\big)}{\partial t}\Delta t + O(\Delta t^2) \\
&\approx \mathbf{R}_{i,j,k}^n\big(\mathbf{w}_{i,j,k}^n\big) + \frac{\partial \mathbf{R}_{i,j,k}^n}{\partial \mathbf{w}_{i,j,k}^n}\big(\mathbf{w}_{i,j,k}^{n+1} - \mathbf{w}_{i,j,k}^n\big). \quad (6.258)
\end{aligned}
$$

Using this linearization in pseudotime, Equation (6.257) becomes a sparse system of linear equations. For the solution of this system, the conjugate gradient method with a simple Jacobi preconditioner is used.

6.5.4.5 Time-Dependent Channel Flow Simulation

In this section, the flow in a square channel is considered. The mean flow direction is the x-direction, while the channel lower and upper walls are placed at $z = 0\sigma$ and $z = 40\sigma$, respectively. The flow is assumed constant in the y-direction. The considered domain is 40σ long in all three coordinate directions. Although the flow is two-dimensional, a three-dimensional solution method is used here, hence the use of the constant y-direction. A finite-volume discretization method is used with a uniform mesh with nine cells in both x- and y-directions, while a stretched mesh with eighteen cells is used in the z-direction. The time-dependent problem starts from a steady flow established by a constant pressure gradient $dp/dx = -0.005$. From time $t = 100$ to $t = 500$ this pressure gradient is then linearly increased to $dp/dx = -0.010$. The time step used in the finite-volume method is $dt = 2$ (macroscale time units). The molecular dynamics method is used in this example to evaluate the viscous stresses on the first four faces of each cell near both domain walls, i.e., the cell face on the solid wall and the first three faces away from the wall. Due to the homogeneity and periodicity of the flow, these four microscale solutions are similarly used for the whole x-y planes for the near-wall cell layers. Due to the symmetry of the problem, these "microscale" viscous are reused for both lower and upper domain walls. The remaining cell faces use a Newtonian fluid assumption for the viscous flux formulation, with the viscosity of the medium being assumed constant at $\mu = 2.0$.

For the idealized case in which all cell faces use Newtonian viscous stresses, the solution of this flow problem is shown as the solid black line in Figure 6.55(a), where the velocity in the center of the domain is plotted versus time.

For the Lennard-Jones fluid considered in the molecular dynamics method, the density is assumed to be $0.80\sigma^{-3}$, while the temperature is $T = 1.50$. Under these conditions, the Lennard-Jones fluid has a viscosity of around 2.03, i.e., very close to the assumed constant value in the Newtonian fluid part of the computational domain. As discussed previously, the viscous stresses as functions of the imposed velocity gradients are computed using the Lees-Edwards boundary conditions. In the molecular dynamics simulations, ensemble as well as temporal averaging is employed. Typically, four or eight independent realizations for each shear rate are constructed, which are then sampled through a sampling duration of typically 100 Lennard-Jones time units. For the MD time step here, i.e., 0.001τ, this involves $100,000$ MD (Molecular Dynamics) time steps.

6.5.4.6 Numerical Experiments: Splitting Methods for Coupled Micro-Macro System of Equations

In this section, we propose a time-integration method derived from the implicit dual-time stepping method and evaluated for the time-dependent channel flow example problem.

Fixed-Interval Microscale Evaluations

In this example, the coupling with the molecular dynamics method takes place by introducing a correction to the Newtonian shear stresses for the cell faces with an associated microscale MD shear stress evaluation, as presented in Equation (6.254). In this method, the following steps are used:

- For time step $<=100$, the solution is marched forward in time using the Newtonian shear stresses evaluation employed throughout the domain.

- At time steps 20, 40, 60, 80, and 100, a microscale MD problems are constructed for the four cell faces nearest to the domain walls. For each cell face, four independent realizations are created, while the viscous stresses are sampled over 100τ after an initial equilibration stage of 50τ.

- For time steps >100, the viscous fluxes in the four cell faces nearest to the domain walls are corrected using an apparent viscosity derived from averaging over the last five MD evaluations for each cell face.

- Every 20 time steps, another set of microscale MD problems are constructed based on the current cell face velocity gradient, and following a new set of MD viscous stress evaluations, the "averaged apparent" viscosity is updated.

The above method is a simple method to take into account the shear-rate dependence of the apparent viscosity, while greatly reducing the computational overhead compared to a full set of MD shear stress evaluations computed for each macroscale time step. However, for rapidly changing macroscale velocity gradient, using the "running" average of the latest MD predictions with a number of previous evaluations, introduced a potential time lag in incorporating the shear-rate dependence of the apparent viscosity.

The method can be summarized as follows:

Algorithm 6.45. *On a uniform time grid with $t^n = t_0 + n\Delta t$, $n = 0, \ldots, N$, (where N is given), the discretized coupled macro- and microscale equations are integrated in time from time level n to $n+1$ using the following scheme:*

1) *initialize the averaged apparent viscosity $\mu_{ave} = \mu_{approx} = 2.0$*

2) *if $n < 100$ and n is multiple of $n_{interval}$ go to 3) else go to 4), (here, $n_{interval} = 20$)*

3) *dual-time step update based on fixed μ_{ave}*

    i) for the cell faces with microscale fluxes, compute the velocity gradients

    ii) construct the viscous flux corrections f_i using updated apparent viscosity

    iii) perform dual-time step update using n_{Newton} relaxation steps (here, $n_{Newton} = 25$)

4) *dual-time step update based on updated μ_{ave}*

    i) for the cell faces with microscale fluxes, compute the velocity gradients

   ii) initialize the molecular-dynamics microscale problems with the imposed velocity gradients from the finite-volume cell faces and integrate these through the initial equilibration phase (e.g., $t_{equi} = 50\tau$)

   iii) integrate microscale problems in time through t_{sample} microscale time and average apparent viscosity in time and ensemble average over $n_{ensemble}$ independent realizations

   iv) compute the new μ_{ave} as the average over the last n_{window} (including present) molecular dynamics solutions (here $n_{window} = 5$)

   v) construct the viscous flux corrections f_i using updated apparent viscosity

   vi) perform dual-time step update using n_{Newton} relaxation steps

if $n < N$ go to 1)

Figure 6.55 presents the results for the velocity predicted with the above method from two independent realizations. Compared to the idealized case with Newtonian fluxes throughout the domain, the use of MD microscale fluxes leads to a small reduction in the cell center velocity, since the apparent viscosity predicted by the MD simulations is slightly higher than the constant viscosity used in the remainder of the domain. Also, using the averaging over five evaluations, the statistical scatter in the MD microscale fluxes only leads to modest fluctuations in the macroscale velocity field compared to the fully Newtonian case.

The predicted apparent viscosity and the resulting "running averages" are presented in Figure 6.56, clearly showing the significant reduction of the statistical scatter in the MD data when such an averaging is used in addition to the already used temporal and ensemble averaging.

The present dual-time stepping method solves a "quasi-steady" problem at each time step. For each of these "quasi-steady" problems, an implicit solution method is used based on an underrelaxed Newtonian relaxation process. The convergence of this implicit system for a number of time steps is presented in Figure 6.57. In the examples shown, twenty-five "pseudosteps" are used, leading to a reduction of the maximum norm of the residual (which now includes the "unsteady" flow contribution) of at least six orders of magnitude. When this residual norm is reduced to machine precision, the only discretization errors due to the time discretization are those due to the truncation errors in the employed implicit three-point stencil. With a reduction of the residual norm of six or seven orders of magnitude, it can be expected that the additional contributions are much smaller than the truncation errors in the implicit three-point stencil.

Remark 6.46. *The iterative splitting method is used as a coupling scheme to embed a small scale problem (molecular dynamics equation) to a large scale problem (Navier-Stokes equation). To obtain a computable method, we employ different ideas to reduce the computational time, e.g., fixed microscale evaluation. More time-consuming algorithms are discussed in [94].*

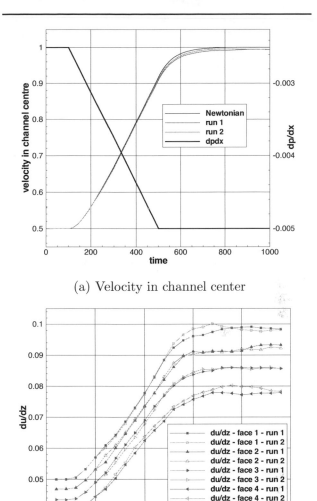

(a) Velocity in channel center

(b) Velocity gradients near walls

FIGURE 6.55: Velocity in channel center (figure (a)), velocity gradients near walls (figure (b)). Dual-time stepping method with averaging of apparent viscosity. Finite-volume discretization method with molecular dynamics viscous fluxes on first four cells near lower and upper walls. Lennard-Jones fluid at $0.80\sigma^{-3}$ and $T = 1.50$. MD data averaged over four realizations, $\tau_{sample} = 100$.

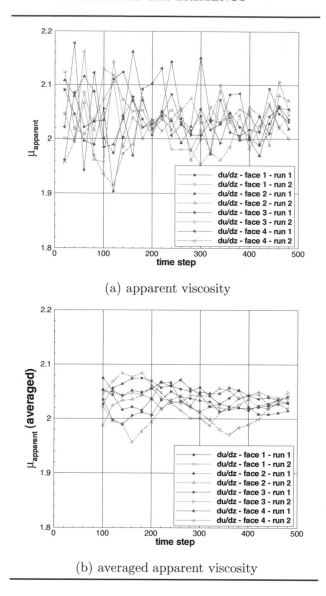

(a) apparent viscosity

(b) averaged apparent viscosity

FIGURE 6.56: Apparent viscosity (figure (a)), averaged apparent viscosity (figure (b)). Dual-time stepping method with averaging of apparent viscosity. Finite-volume discretization method with molecular dynamics viscous fluxes on first four cells near lower and upper walls. Lennard-Jones fluid at $0.80\sigma^{-3}$ and $T = 1.50$. MD data averaged over four realizations, $\tau_{sample} = 100$.

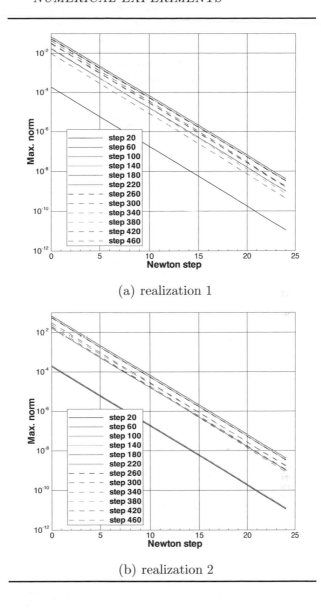

(a) realization 1

(b) realization 2

FIGURE 6.57: Realization 1 (figure (a)), realization 2 (figure (b)). Channel flow with time-dependent pressure gradient. Dual-time stepping method with averaging of apparent viscosity. Convergence of "inner-loop" in dual-time step method. Channel flow with time-dependent pressure gradient. Lennard-Jones fluid at $0.80\sigma^{-3}$ and $T = 1.50$. MD data averaged over four realizations, $\tau_{sample} = 100$.

6.6 Conclusion to Numerical Experiments: Discussion of Some Delicate Problems

In some real-life applications, we have implemented the splitting methods as solver schemes and discussed the results.

In the following, we point out that the following advantages and disadvantages of the iterative splitting scheme can occur to such delicate problems:

- Iterative splitting schemes can be implemented with standard codes, for example, Unstructured Grid (UG), see [15], and MATLAB® codes, e.g., Finite Difference and Operator Splitting Methods (FIDOS). Such implementations to standard codes improve the accuracy of the underlying splitting schemes.

- We achieve higher-order results with more iterative steps also for complicated equations, e.g., multiscale or stiff problems.

- We can strike a balance between iterative steps and time steps, so that we can reduce the computational time and yet achieve sufficient accuracy.

- In the implementation, we can combine stiff and nonstiff operators: while the stiff operators are solved implicitly, the nonstiff operators are solved explicitly.

- Standard implementations can be made more flexible by iterative schemes, while the choice of implicit and explicit operator can be done with different stages.

- One of the delicate problems is the computation of $\exp(At)$ (e.g., if we deal with a time integration method) and the integrals of such terms. To solve such problems, we can use the ideas of the so-called ϕ-functions and Krylov-subspaces, see [131], [130], and [187], to derive simpler and computable operators.

- We can also combine splitting methods with spatial discretization methods and adaptive local discretization in space in [87].

Remark 6.47. *Nevertheless, the benefits of the iterative splitting methods are its flexibility and in the future such iterative schemes can be applied as code coupling methods. The benefits can be present in the following points:*

- *Reduction of the computational time by decomposing into simpler and more easily solved parts.*

- *Adequate scale-dependent discretizations reduce the stiffness in the equations.*

- *Implicit and explicit operators reduce computational time, while the explicit operators are cheap to compute.*

- *Higher-order results can be achieved with more iteration steps or smaller time steps. Balancing such steps, we can obtain the optimal results.*

Chapter 7

Summary and Perspectives

This book discusses the numerical analysis and applications of iterative operator splitting methods for differential equations, based on different time and space scales.

While the numerical analysis of iterative splitting methods is considered for bounded operators, we could also extend the methods to nonlinear, stiff, and spatio-temporal problems in the applications. We discuss applications to real-life problems that are based on parabolic and hyperbolic equations.

The following problems are solved:

- The generalization of consistency and stability results to nonlinear, stiff, and spatially decomposed splitting problems.

- The acceleration of computation by decoupling into simpler problems.

- The efficiency of the decomposed methods.

- This theory is based on semigroup analysis, which is well understood and applicable to the splitting methods.

- Applications to several computational methods, including flow problems, elastic wave propagation, and heat transfer.

Based on iterative splitting methods, there is an important fixed-point problem with respect to the initial solutions or starting solutions.

The following items are discussed in several papers by the author in order to extend iterative splitting methods in different directions:

- Improved starting solution for the first iteration step by solving the fixed-point problem accurately, e.g., using an embedded Newton solver, see [105].

- Parallelization of the iterative methods on the operator level, i.e., a first idea is called the *windowing* (e.g., decomposition in time domains) and is discussed in [91].

- Decoupling algorithms with respect to the spectrum of the operators and automation of the decomposition process, see [93].

- Simulation of real-life problems with improved parallel splitting methods, see first ideas [96].

Perspectives for these decomposition methods include their combining the efficiency of decoupling with a generalization to nonlinear and boundary problems.

In the future, decomposition methods will be adaptive methods that respect the physical constraints of the equation and accelerate the solution process for a complicated system of coupled evolution equations.

Even numerical experiments can point at more solutions, and ones adequate for multiphysics applications, and hence such results can be useful even without analytically exact solutions of the complicated equations involved.

Our solutions offer the possibility of mathematically exact proofs of existence and uniqueness of solutions of parabolic differential equations, whereas the older proofs of greater complexity lack these convergence and existence results. So we would like to investigate more to see the application of the iterative splitting methods proposed in this book.

We have deferred to mathematical correctness when there is a chance to fulfill this in the simpler equations, but we have also described very complex models and show their solvability without proofs of existence and uniqueness. To find a balance between simple provable equations and complex calculable equations, we present iterative splitting methods for decoupling complex equations into provable equations. We consider the underlying theory to systems of ordinary differential equations and could simplify the theoretical aspects to boundable operators, such that we could extend the theory with respect to boundable spatial discretized partial differential equations, see [111].

Chapter 8

Software Tools

8.1 Software Package Unstructured Grids (UG) (P. Bastian et.al. [15])

In this section we introduce the software package $\mathbf{r^3t}$. The real-life problems of waste disposal scenarios are integrated in this package and all numerical calculations were performed with it.

This software package was developed for the task of a project to simulate transport and reaction of radioactive pollutants in groundwater. The specifications for the software package were flexible inputs and outputs as well as the use of large time steps and coarse grids to achieve the claimed long time calculations.

The concept of the Unstructured Grids **(UG)** software toolbox and the different software packages **UG**, $\mathbf{d^3f}$, $\mathbf{r^3t}$, and GRAphics Programming Environment **(GRAPE)** are to the fore.

For satisfying these tasks, we built on the preliminary **UG** software toolbox that was developed at the institute. We use the name *software toolbox* to indicate that the software tools can be provided for other software packages. The software shall not be conclusive in the sense of a software package, but be extendable. The development consisted of compiling a flexible input interface. Furthermore, a coupling concept was developed that uses data from other software packages. More particularly, the implementation of efficient discretizations and error estimators was able to achieve the claimed large time steps and coarse grids for the numerical calculations.

The main topics of this chapter consist of the rough structuring of the software packages, the description of the **UG** software toolbox, the description of the $\mathbf{d^3f}$ software package, the description of the $\mathbf{r^3t}$ software package, as well as their concepts.

8.1.1 Rough Structuring of the Software Packages

Software Toolbox **UG**: The **UG** software toolbox manages unstructured, locally refined multigrid hierarchies for 2D and 3D geometries. The software toolbox provides the software structures for the further modules. It has a variety of usable libraries for discretizations and solutions of systems of partial differential equations, as well as interactive graphics to display the results of simulations. Therefore, it serves as a basic module for the development and programming of the $\mathbf{r^3t}$ and $\mathbf{d^3f}$ software packages.

Software Package $\mathbf{d^3f}$: This package was developed to solve densely floated fluxes of groundwater. It treats systems that consist of a flux equation and a transport equation, which are nonlinear and coupled. The simulation calculations can treat steady as well as unsteady fluxes. The software package has a flexible input setting and yields velocity data for the transport calculations of the $\mathbf{r^3t}$ software package.

Software Package $\mathbf{r^3t}$: This software package was developed for solving transport-reaction equations of various species in groundwater flowing through porous media. For the solutions of convection-dominant equations, an improvement of the discretization was developed. The goal was to be able to use large time steps and coarse grids, so as to be able to accomplish the calculations within a reasonable and specifiable time frame, and yet be able to make predictions over a long model time period. The error estimators and solvers were adapted to these finite volume discretizations. The flexible input of parameters for the model equation as well as the output of the data of solution and grid are further tasks, that the software has to fulfill.

Software Package **GRAPE**: This software package is a wide-ranging visualization software, that is particularly suited for visualizations of solutions on unstructured grids with various presentation methods. The results of the $\mathbf{r^3t}$ module are visualized using this package. These wide-ranging visualization capacities of **GRAPE**, see [116], could be used to gain an improved comprehension of the simulation results.

8.1.2   UG Software Toolbox

The basis of all subsequently described software packages is the **UG** software toolbox. It is derived from the words "<u>U</u>nstructed<u>G</u>rids" and is a software toolbox for the solving system of linear or nonlinear partial differential equations.

The concept of the **UG** software toolbox is the management of an unstructured, locally refined multigrid hierarchy for 2D and 3D grids.

This software package was developed in the beginning of the 1990s, see [15], using the adaptivity of grids, multigrid methods, and parallelization of methods to solve systems of partial differential equations. Because of its effective and parallel algorithms, the software package has the capacity to solve wide-ranging model problems.

Several works accrued in the fields of numerical modeling and numerical simulation of linear and nonlinear partial differential equations: the mechanics of elasticity, densely floated fluxes, transport of pollutants through porous media, and the mechanics of fluxes. Some examples can be found in [14], [15], [141], and [142].

Subsequently, we introduce some applications of the **UG** software toolbox.

For the application of the software toolbox **UG** in a software package, e.g., r^3t, we distinguish between two modules when structuring the emerging software package.

ug: This module is the part of the software package that is independent of the application. It contains software tools such as the interpreter, the event handler, the graphics, the numerics, the local grid hierarchy, the description of areas, the device connection, as well as the storage management. The lowest level contains the parallelization of the program using a graph-based parallelization concept. These tools can be applied to further parts of the application-oriented area.

r3t: In this module the model problem with its appropriate equation is prepared for the application. For its implementation, the problem class library and the application were developed. The module depends on the application in contrast to the **ug** module, that is independent of the application.
The following classification was constituted: Discretizations depending on the problem as well as the according error indicators, the sources and sinks, and the analytical solutions, that were developed especially for these systems of differential equations, are implemented in the program class library.

For the discretization, the finite volume method of higher order was used, see [68] and [88]. As error indicators for the convection-diffusion-dispersion equation, the *a posteriori* error indicators were used, see [178].

The applications contain the initial and boundary conditions as well as the coefficient functions for the equation. Further implemented are the flexible input of equation parameters and several time-independent source terms. In addition, script files are included in the application, these are loaded by the **UG** command interpreter and control the sequence of calculations. Especially, the scanner and parser module were programmed and implemented into the application. The number of equations, phases (with transport and reaction parameters), sources, the boundary conditions, the area parameter, and the hydro-geological parameters, are set at the cue of the program. All files that are used for the input of velocity and geometry data, as well as the output of solution and grid data, can be found in the application.

8.1.3 UG Concept

Essentially, two concepts of the **UG** software toolbox were also used for the software package r^3t. One concept is the unstructured grid concept, that allows the managing of relatively complex geometries with adaptive operations to make the hierarchic methods applicable.

A further concept is that of sparse matrices. It allows the block-by-block savings of wide-ranging systems of partial differential equations to achieve an effective storage and velocity.

These two concepts are decisive for achieving the required efficiency of the r^3t software package and are described subsequently.

We first describe the concept of unstructured grids.

The basis for the **UG** software toolbox consists of the creation and modification of unstructured grids. Using unstructured grids, complex geometries can be approximated.

An efficient structure of the grid data is necessary to produce grids and perform local refinements as well as coarsenings. The geometrical data structure administers geometrical objects, as for example elements, edges, and nodes. The algebraic data structure administers matrices and vectors.

The grids consist of triangles and quadrangles in 2D, or tetrahedrons, pyramids, hexahedrons, and prisms. The structure of the grid data is constructed hierarchic and the elements can be locally refined and coarsed.

The application of multigrid methods with varying smoothing processes is possible because of the locally refineable areas.

The solutions on higher grid levels are only considered on the refined elements. Hence, local grids are treated with all refined elements and some copied elements. One obtains a high efficiency in processing.

A further important concept for the efficiency of the calculations with the r^3t software package is the saving of the sparse matrices that are used in the **UG** software toolbox.

Sparse matrices occur when grids are saved. Based on the local discretizations, only the neighbors are used for the calculations, hence only some entries appear outside of the diagonals. Sparse matrices also occur in the application of weakly coupled systems of partial differential equations. The connections between the individual equations are saved as entries outside of the diagonal. Due to the weak coupling only a few entries appear.

The sparse matrix storage method [172], was developed in order to avoid an ineffective saving of fully populated matrices. It was chosen for this software since it is an efficient storage concept. The idea of this method is based on the saving of sparse matrices in a block-matrix graph, in which blocks are indicated as local arrays over a set of compact row-orientated patterns, see [172]. This combines the flexibility of graph structures with the higher efficiency based on higher data densities. The compact inner pattern allows the identification of entries, and with that, further advantages of saving and calculation times can be obtained. The compact storage technique is row- or column-oriented; preliminary individual matrix entries can be replaced by block entries that represent a system of partial differential equations. The entry in the diagonal of the block matrix is the coupling within the equations, the entry outside the diagonal of the block matrix is the coupling between neighboring nodes, if a space term is used. The saving as vector-matrix graph is considered in Figure 8.1.

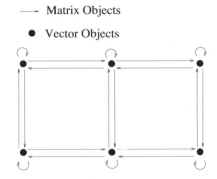

FIGURE 8.1: Vector-matrix graph for efficient saving.

Hence, the claimed conditions to solve wide-ranging systems of partial differential equations on complex geometries were satisfied.

8.1.4   Software Package d³f

The previous project about the **r³t** software package was posed by the Society for Plant Safety and Reactor Safety (Gesellschaft für Anlagen- und Reaktorsicherheit mbH, GRS) in Braunschweig. It was developed to simulate densely floated fluxes of groundwater. It contains an implementation of the catalogues for problem classes and applications. In the problem class, error estimators, solvers, and discretizations for the equations were implemented. The catalogue for applications contains several applications. Particularly implemented are special boundary and initial conditions, special prefactors for the equations, as well as sources and sinks. The initialization of the equation parameters, as well as the boundary and the initial conditions, and the parameters for sources and sinks are read in from files at start time. Therefore, flexible calculations can be made in 2D and 3D model examples.

8.1.5   Equations in d³f

Two coupled nonlinear time-dependent partial differential equations were implemented in the software package **d³f**. These equations are subsequently considered.

The equation for the flux is given as:

$$\partial_t(\phi\rho) + \nabla \cdot (\rho\mathbf{v}) = Q, \tag{8.1}$$

where ρ denotes the density of the fluid, $\mathbf{v}$ calculates the velocity using Darcy's law, ϕ is the effective porosity, and Q stands for the source term.

We use Darcy's law to derive the velocity of the fluid through the porous media. Darcy's law is a linear relationship between the discharge rate through a porous medium, the viscosity of the fluid, and the pressure drop over a given distance:

$$Q = \frac{-\kappa A}{\mu} \frac{P_b - P_a}{L}. \tag{8.2}$$

The total discharge, Q (units of volume per time), is equal to the product of the permeability (κ units of area) of the medium, the cross-sectional area A of flow, and the pressure drop $P_b - P_a$, all divided by the dynamic viscosity μ, and the length L of the pressure drop. Dividing both sides of the equation by the area, we obtain the general notation:

$$q = \frac{-\kappa}{\mu} \nabla P, \tag{8.3}$$

where q is the flux and ∇P is the pressure gradient vector. This value of flux is often called the *Darcy flux*.

The pore velocity v is related to the Darcy flux q by the porosity ϕ and is given as:

$$v = \frac{q}{\phi}. \tag{8.4}$$

The pore velocity is used as the velocity of a conservative tracer and as our velocity for the simulations given by physical experiments.

The transport equation is given as:

$$\partial_t(\phi \rho c) + \nabla \cdot (\rho \mathbf{v} c - \rho D \nabla c) = Q', \tag{8.5}$$

where c is the concentration of the solute, D denotes the diffusion-dispersion tensor, and Q' is the source term.

The first Equation (8.1) calculates the flux through a porous media using Darcy's law as a continuity equation and gives the density of the fluid. The second equation is a transport equation for water in dissolved salt. It is a time-dependent convection-diffusion-dispersion equation, where the convective factor is calculated using Darcy's law. Depending on the variation in the density, vortical fluxes can occur. Based on the small diffusions in water one obtains a convection-dominated equation, for this equation adapted solver and discretization methods were developed, see [141] and [142].

8.1.6   Structure of d³f

For inputting equation and control parameters, a preprocessor was developed. The preprocessor reads the wide-ranging input parameters before the calculations start. The essential module and centerpiece is the processor, which was developed on the basis of **UG** and adapted with further problem-specific classes for the intended equations. Applications that contain the parameter files and control files for program runs are filed in a further library. For data to be circulated, a postprocessor was developed, which presents an interface and saves the data of the solutions in files. As a result other programs, especially the visualization software **GRAPE**, can read the data and reprocess them.

8.2 Software Package r³t

In this section we present, in a more or less rough fashion, the software package **r³t**: the name is derived from "**R**adionuclide, **R**eaction, **R**etardation and **T**ransport" (**r³t**). This software package discretizes and solves systems of convection-diffusion-dispersion-reaction equations with equilibrium sorption and kinetic sorption.

8.2.1 Equation in r³t

The equation for equilibrium-sorption is given by:

$$\phi \, \partial_t \, R_i \, c_i + \nabla \cdot (\mathbf{v}c_i - D\nabla c_i) \qquad (8.6)$$
$$= -\phi \, R_i \, \lambda_i c_i + \sum_{k=k(i)} \phi \, R_k \, \lambda_k c_k + \tilde{Q}_i,$$

$$c_{e(i)} = \sum_i c_i,$$

$$R_i = 1 + \frac{(1-\phi)}{\phi} \rho \, K(c_{e(i)}),$$

$$\text{with} \quad i = 1, \dots, M,$$

where c_i denotes the i-th concentration, R_i the i-th retardation factor, λ_i the i-th decay factor, and $\mathbf{v}$ is the velocity, which is either calculated in the **d³f** software package or preset. Q_i denotes the i-th source term, ϕ the porosity, $c_{e(i)}$ the sum of all isotope concentrations referring to element e, and K is the function of the isotherms, see [69].

8.2.2 Task of r³t

The software package was commissioned by the GRS for prognoses in the area of radioactive repositories in salt deposits.

The following properties were specified for the development of the software package **r³t**:

- Independence of the transport equation, that has to be solved, shall be achieved. This can be obtained by applying a 2D or 3D version based on the flexible concept.

- A simple adaptability on the shell level shall be possible. A script file with preferences for the parameters of discretization and the solver, as well as a graphical visualization shall be used. A suitable definition of interfaces for the output of data for reprocessing by a graphic software as, e.g., **GRAPE**, cf. [116], shall be done.

- The programming of self-contained modules shall be developed, the modules shall be structured with accurately defined interfaces. Therefore, several programmers can be involved in the development.

- The differential equations shall be applied as scalars and also as systems. The coupling of at least four phases shall be possible.

- All procedures shall be possible for a simultaneous computer to allow wide-ranging calculations.

The structure of the software package r^3t is based on the preliminary work of the developed software toolbox **UG**, see Section 8.1.2, and was advanced in this context. The presented tasks were realized using the following concepts.

8.2.3   Conception of r^3t

The software conception of r^3t is given with the following approaches.

While developing our software package r^3t, the philosophy and conception of the *UG* development was further evolved. User-independent numerical algorithms were programmed and provided for special applications. During the development of the software package r^3t, further numerical algorithms concerning discretizations, solvers, and error indicators were implemented. The used application-independent structures, the so-called *numerical procedures*, were continued, confer [142].

The essential renewals of this development phase consisted of connecting the existing software packages and combining the corresponding tasks of the respective programs. Thus, the software package d^3f, which is used to calculate the flux of the transport equation, [58] and [141], was coupled with the software package r^3t. A coordinate-oriented and grid-independent storage of the velocity data was developed. A further connection to the visualization software **GRAPE**, cf. [186], could be used to visualize the results.

A flexible input was developed for the wide-ranging parameters that are available in the form of input files. Hence, wide-ranging test calculations are possible, as is necessary to meet the specifications. This further flexibility is adequate for the concept of modular programming. Individual parts of the program, such as the preprocessor and the corresponding discretization and solution methods, can be developed independently. A further flexibility is obtained from the data concept, which enables the input and output of files using interfaces, such that the individual software packages were coupled.

The application of the software package $\mathbf{r^3t}$ with its connections to other software packages is described next.

8.2.4   Application of $\mathbf{r^3t}$

The application of $\mathbf{r^3t}$ is outlined in Figure 8.2.

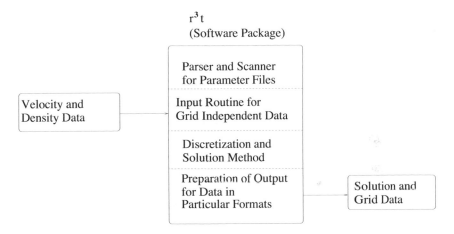

FIGURE 8.2: Application of $\mathbf{r^3t}$.

The approach of software package $\mathbf{r^3t}$ illustrated in Figure 8.2 is now explained.

The preparations up to the point where the model problems are calculated consist of the input of equation parameters and control files. In particular, the number of equations and phases as well as the parameters for the convection-diffusion-dispersion-reaction equations can be ascertained. These are read in during the start of the initialization phase, as, e.g., the data of geometry and velocity, that are grid-independently stored in files by the software package $\mathbf{d^3f}$.

The actual calculation phase of the software package proceeds after the initialization. Problem-specific discretization as well as the solvers for linear equation systems, are used. Error indicators are used for adaptivity to realize effective calculations in time of equations with local grid refinements. Especially velocity and density data are used for adaptive calculations. The individual subequations, that were treated using various discretizations, see [69], are coupled afterwards with operator-splitting methods.

The results of the calculation steps are read out into a particular file format in the output phase, and are processed afterwards by the postprocessor **GRAPE**. During the preparations, the total concentration is calculated by using the explicit correlation of mobile and adsorbed concentration, see Equation (1.1), for the case of equilibrium-sorption. The associated grid data belong to the output data of solutions and describe the geometric proportions of the data.

Hence, the application of the r^3t software package was integrated within the ambit of further software packages that realize preprocessing or postprocessing.

We next describe a coupling concept that achieves a flexible interchange of files between the software packages involved.

8.2.5 Coupling Concept of r^3t

The concept of coupling between individual software packages is now developed based on interchanges of files at predefined interfaces. This connection is sufficient for parameters that are only insignificantly changed during the time-dependent course and have no backcoupling on other programs.

The problem-dependent parameters, that are constant during the calculation time, can be read in at the beginning, e.g., the geology of rocks in the area, the exchange parameters of radioactive species, as well as the determination of sources and sinks.

Further inputs can vary in the course of the simulation time, e.g., velocity or density data. These are read in during the operation, at their specific time points. The data are coupled single-sided to the r^3t software package, i.e., the results of r^3t are not used for the calculation. The reason is the modeling, where the radioactive contaminates have no backcoupling on the flow field. The backcoupling on software packages d^3f therefore can be omitted.

Completed calculations are saved in files and for further utilization read in from the postprocessor.

A strong coupling between the two programs was not pursued any further. This is only advisable for a coupling between flux and contaminants, then transport and flux equation are solved together.

Using the software package r^3t, the subsequent model examples can be realized and evaluated.

8.3 Solving Partial Differential Equations Using FIDOS (authors: J. Geiser, L. Noack)

This section describes the PDEs treated, the methods used, and the usage of the program package FIDOS 1.0 (Finite Difference and Operator Splitting Methods).

The first version, FIDOS 1.0, covers only computations, that already have been published and used in this monograph. In the following subsections we discuss the treatment of the methods, the assembling, and the programming structure.

8.3.1 PDEs Treated

The PDEs treated in this paper have the form

$$D_t u = D(u)u, \tag{8.7}$$

where the derivative in time is of the first or second order,

$$D_t u = \partial_t u \text{ or } D_t u = \partial_{tt} u,$$

and the spatial operator $D(u)$ can possess derivatives in from one to three spatial dimensions, depending on the PDE:

$$D(u)u = D_x(u)u + D_y(u)u + D_z(u)u.$$

The spatial operators can consist of linear terms, e.g.,

$$D_x(u) = A\partial_x u \text{ or } D_x(u) = A\partial_{xx} u,$$

and can contain nonlinear terms, as well,

$$D_x(u) = Au\partial_x u,$$

where A (or B, C, respectively) is a constant.

The types of PDEs that can be examined using FIDOS therefore varies from simple wave and heat equations to nonlinear momentum equations. The package can easily be extended to other kinds of PDEs and more methods. The PDEs of the first version of the program package are listed below.

8.3.1.1 Wave Equation

The wave equation for three dimensions is given by

$$\partial_{tt} u = A\partial_{xx} u + B\partial_{yy} u + C\partial_{zz} u, \text{ with} \tag{8.8}$$

$$u(x, y, z, 0) = \sin\left(\frac{1}{\sqrt{A}}\pi x\right) \sin\left(\frac{1}{\sqrt{B}}\pi y\right) \sin\left(\frac{1}{\sqrt{C}}\pi z\right), \tag{8.9}$$

$$\partial_t u(x, y, z, 0) = 0. \tag{8.10}$$

The analytical solution in $\Omega = [0, 1]^3$ is given by

$$u(x, y, z, t) = \sin\left(\frac{1}{\sqrt{A}}\pi x\right) \sin\left(\frac{1}{\sqrt{B}}\pi y\right) \sin\left(\frac{1}{\sqrt{C}}\pi z\right) \cos\left(\sqrt{3}\pi t\right).$$

The one- and two-dimensional case follow analogously, whereby the constant $\sqrt{3} = \sqrt{d}$ has to be changed according to the dimension d.

The wave equation can be treated for a nonstiff case, e.g., $A = B = C = 1$, as well as for stiff PDEs, e.g., $A = 0.01, B = 1, C = 100$. See [78, 77] for further explanations.

8.3.1.2 Viscous Burgers Equation

The viscous Burgers equation is given as

$$\partial_t u = -u\partial_x u - u\partial_y u + \mu(\partial_{xx} u + \partial_{yy} u), \tag{8.11}$$

where the first two terms of the right-hand side together are the nonlinear operator $A(u)u$, whereas the last term can be regarded as a linear operator Bu.

The analytical solution of this PDE is

$$u(x, y, t) = \left(1 + e^{(x+y-t)/2\mu}\right)^{-1}.$$

8.3.1.3 Mixed Convection-Diffusion and Burgers Equation

The mixed convection-diffusion and Burgers equation is

$$\partial_t u = -\frac{1}{2}(u\partial_x u + u\partial_y u + \partial_x u + \partial_y u) + \mu(\partial_{xx} u + \partial_{yy} u) + f(x, y, t), \quad (8.12)$$

here the analytical solution is chosen to be

$$u(x, y, t) = \left(1 + e^{(x+y-t)/2\mu}\right)^{-1} + e^{(x+y-t)/2\mu}.$$

The function f is calculated accordingly. The equation is again split into a nonlinear and a linear part to obtain the operators $A(u)u$ and Bu.

8.3.1.4 Momentum Equation

The momentum equation is given as

$$\partial_t \mathbf{u} = -\mathbf{u} \cdot \nabla \mathbf{u} + 2\mu\nabla(\mathbf{u} \cdot \mathbf{u} + \mathbf{u} \cdot \mathbf{v} + 1/3\nabla \mathbf{u}) + \mathbf{f}(x, y, t), \tag{8.13}$$

with analytical solution $\mathbf{u} = (u_1, u_2)^T$ and

$$u_1(x, y, t) = u_2(x, y, t) = \left(1 + e^{(x+y-t)/2\mu}\right)^{-1} + e^{(x+y-t)/2\mu}.$$

The function f is again calculated accordingly. The one-dimensional formulation follows analogously, see [83].

8.3.1.5 Diffusion Equation

Being a diffusion equation, the two-dimensional heat equation is treated as an example. These methods can easily be applied to any other diffusion equations.

$$\partial_t u = A\partial_{xx} u + B\partial_{yy} u, \tag{8.14}$$

with analytical solution

$$u(x, y, t) = e^{-(A+B)\pi^2 t} \sin(\pi x) \sin(\pi y).$$

8.3.2 Methods

In general, all methods used, independently of their explicit or implicit character and independently of the number of needed substeps, satisfy

$$u^{n+1}(x, y, z, t) = METHOD(u^n(x, y, z, t), u^{n-1}(x, y, z, t)) \tag{8.15}$$

with given initial condition

$$u^0(x, y, z) \text{ for } t = 0, \tag{8.16}$$

and if needed

$$u^1(x, y, z) \text{ for } t = 0, \tag{8.17}$$

if the space consists of three dimensions, analogous formulations hold for less dimensions. In the following, the number of dimensions is always set to be two, if not mentioned otherwise. All methods can be applied to one or three spatial dimensions, too.

8.3.2.1 ADI Method

The ADI method is given next, for two-dimensional equations with first-order derivatives in time.

The Crank-Nicolson method,

$$\left(L_1 L_2 + \frac{\Delta t}{2}\left(L_2 A\partial_{xx} + L_1 B\partial_{yy}\right)\right) u^{n+1}$$

$$= \left(L_1 L_2 - \frac{\Delta t}{2}\left(L_2 A\partial_{xx} + L_1 B\partial_{yy}\right)\right) u^n \tag{8.18}$$

with operators L_1, L_2 split into two substeps, yielding the ADI method,

$$(L_1 + \Delta t/2 \, A\partial_{xx})u^{n+1,1} = (L_1 - \Delta t/2 \, A\partial_{xx})(L_2 - \Delta t/2 B \, \partial_{yy})u^n, \tag{8.19}$$

$$(L_2 + \Delta t/2 \, B\partial_{yy})u^{n+1} = u^{n+1,1}. \tag{8.20}$$

Depending on the operators L_1, L_2, the ADI method is a second- or fourth-order method. The first case is obtained with $L_1 = L_2 = 1$, the second one with $L_1 = 1 + \Delta x^2/12 \, \partial_{xx}$ and $L_2 = 1 + \Delta y^2/12 \, \partial_{yy}$.

8.3.2.2 LOD Method

The LOD method is derived for partial differential equations with two spatial dimensions, linear operators $D_x u$, $D_y u$, and first- and second-order derivatives in time,

$$
\begin{aligned}
\partial_t u &= (D_x + D_y)u, \\
\partial_{tt} u &= (D_x + D_y)u.
\end{aligned}
$$

Using the finite difference schemes

$$
\partial_t u^n \approx \frac{u^{n+1} - u^n}{\Delta t} \text{ and} \tag{8.21}
$$

$$
\partial_{tt} u^n \approx \frac{u^{n+1} - 2u^n + u^{n-1}}{\Delta t^2}, \tag{8.22}
$$

the LOD method is derived from the explicit Euler method,

$$
\begin{aligned}
u^{n+1} - u^n &= \Delta t(D_x + D_y)u^n, \\
u^{n+1} - 2u^n + u^{n-1} &= \Delta t^2(D_x + D_y)u^n.
\end{aligned}
$$

The LOD method for PDEs with first-order derivatives in time is now given by

$$
\begin{aligned}
u^{n+1,0} - u^n &= \Delta t(D_x + D_y)u^n, \\
u^{n+1,1} - u^{n+1,0} &= \Delta t \eta D_x(u^{n+1,1} - u^n), \\
u^{n+1} - u^{n+1,1} &= \Delta t \eta D_y(u^{n+1} - u^n),
\end{aligned}
$$

while the LOD method for second-order derivatives is

$$
\begin{aligned}
u^{n+1,0} - 2u^n + u^{n-1} &= \Delta t^2(D_x + D_y)u^n, \\
u^{n+1,1} - u^{n+1,0} &= \Delta t^2 \eta D_x(u^{n+1,1} - 2u^n + u^{n-1}), \\
u^{n+1} - u^{n+1,1} &= \Delta t^2 \eta D_y(u^{n+1} - 2u^n + u^{n-1}).
\end{aligned}
$$

8.3.3 Iterative Operator Splitting Methods

Iterative methods repeat the solving steps until a given break criterion is achieved. Such a criterion can be the number of iteration steps (as is used in FIDOS) or an error tolerance. Then the general scheme for an iterative method is

$$
u^{n+1,i} = METHOD(u^{n+1,i-1}), \; i = 1 \ldots m \tag{8.23}
$$

$$
u^{n+1} = u^{n+1,m}. \tag{8.24}
$$

There are different possibilities to choose the initial value $u^{n+1,0}$; all are integrated in FIDOS:

$$
\begin{aligned}
&(1) \quad u^{n+1,0} = u(x, y, t^{n+1}), \\
&(2) \quad u^{n+1,0} = 0, \\
&(3) \quad u^{n+1,0} = u^n.
\end{aligned} \tag{8.25}
$$

If two initial values are needed, they are set to be equal,

$$u^{n+1,1} = u^{n+1,0}.$$

8.3.3.1 Standard IOS Method

The standard iterative operator-splitting method is given with the following scheme, where the partial derivatives are exchanged by using Equation (8.21).

The IOS method for equations with first-order derivatives in time is given as

$$\partial_t u^{n+1,i} = D_x(u^{n+1,i-1})u^{n+1,i} + D_y(u^{n+1,i-1})u^{n+1,i-1}, \quad (8.26)$$
$$\partial_t u^{n+1,i+1} = D_x(u^{n+1,i-1})u^{n+1,i} + D_y(u^{n+1,i-1})u^{n+1,i+1} \quad (8.27)$$

for $i = 1, 3, \ldots, 2m + 1$.

8.3.3.2 Coupled η-IOS Method

For the wave equation, the second-order derivatives in time are replaced using the finite difference scheme (8.22). This method is explicit, but the coupling of the current time step t^{n+1} with the two previous time steps using the η-method is much more effective. The η-method uses the approximation

$$u^n \approx \eta u^{n+1} + (1 - 2\eta)u^n + \eta u^{n-1}$$

for $\eta \in [0, 1/2]$. Equality holds for $\eta = 0$.

Then the scheme of the coupled η-IOS method is given by

$$\begin{aligned}
u^{n+1,i} - 2u^n + u^{n-1} &= \Delta t^2 A(\eta u^{n+1,i} + (1 - 2\eta)u^n + \eta u^{n-1}) \quad (8.28) \\
&+ \Delta t^2 B(\eta u^{n+1,i-1} + (1 - 2\eta)u^n + \eta u^{n-1}), \\
u^{n+1,i+1} - 2u^n + u^{n-1} &= \Delta t^2 A(\eta u^{n+1,i} + (1 - 2\eta)u^n + \eta u^{n-1}) \quad (8.29) \\
&+ \Delta t^2 B(\eta u^{n+1,i+1} + (1 - 2\eta)u^n + \eta u^{n-1}),
\end{aligned}$$

for constants A, B.

8.3.4 Eigenvalue Methods

Eigenvalue methods use the same steps as the standard methods. The difference consists in first determining the stiff and nonstiff operator of the PDE. In the difference schemes the stiff operator is then treated implicitly in the first step, and explicitly in the second step. See [84] for further explanations.

8.3.5   Numerical Examples

When starting the program by typing

```
>> startFIDOS;
```

the user is asked to type in the number(s) of the problem as well as of the
solver(s). Some problems or solvers can be started together. Then the number
of time steps, or a vector of numbers, as well as the number of space steps, or
a vector as well, have to be specified.

When solving a problem for different numbers of time and space steps, it
is interesting to know the numerical rate of convergence,

$$\rho(\Delta t_1, \Delta t_2) = \frac{\ln(err_{L_1}(\Delta t_1) - err_{L_1}(\Delta t_2))}{\ln(\Delta t_1 - \Delta t_2)}. \tag{8.30}$$

FIDOS calculates this rate automatically, if desired, and gives the whole
L_AT_EX-table to be included in a paper with all results and convergence rates
as well as label and caption. One program run is given as an example, where
Table 8.1 and Figure 8.3 is created automatically and includes the results.

```
>> startFIDOS;
-----------------------------------------------------------------------
 FIDOS - solving PDEs with FInite Difference schemes and
 Operator-Splitting methods
-----------------------------------------------------------------------


Problem types:

    [1] wave equation
    [2] viscous Burgers equation
    [3] mixed convection-diffusion Burgers equation
    [4] momentum equation
For more information about the problems, type help,
problem and number of the problem, e.g., >> help problem1

ATTENTION, some methods and/or problems cannot be started together!

    [1] ADI      for problem  1
    [2] LOD      for problems 1
    [3] IOS      for problems 2,3,4
    [4] EVIOS    for problem  2
    [5] etaIOS   for problem  1
    [6] etaEVIOS for problem  2
For more information about the solvers, type help and name of the metho
e.g., >> help ADI
```

Now the specifications of the problem, solver, time steps, space steps, L_AT_EX-table, and plots follow:

```
--------------------------------------------------------
Get problems, solvers and problem-dependent values
--------------------------------------------------------


Problem number(s): 1
Solver number(s): 5

Number(s) of temporal steps on [0,T], e.g., 16 or [4,8]: [4,8,16]
Number(s) of spatial steps on [0,1]^d, e.g., 4 or [4,8]: [4,8,16]

Type 1 to see the tex-tables, otherwise 0: 1
Type 1 to see the plots, otherwise 0: 1
```

The output follows immediately:

```
--------------------
Start Calculations
--------------------


etaIOS, wave, dx=0.25, dy=0.25, dt=0.3125
Errors at time T: L1-error = 0.048967, max. error = 0.13442

etaIOS, wave, dx=0.125, dy=0.125, dt=0.3125
Errors at time T: L1-error = 0.021581, max. error = 0.054647

etaIOS, wave, dx=0.0625, dy=0.0625, dt=0.3125
Errors at time T: L1-error = 0.014382, max. error = 0.035715

etaIOS, wave, dx=0.25, dy=0.25, dt=0.15625
Errors at time T: L1-error = 0.042097, max. error = 0.11556

etaIOS, wave, dx=0.125, dy=0.125, dt=0.15625
Errors at time T: L1-error = 0.014849, max. error = 0.037601

etaIOS, wave, dx=0.0625, dy=0.0625, dt=0.15625
Errors at time T: L1-error = 0.0077193, max. error = 0.01917

etaIOS, wave, dx=0.25, dy=0.25, dt=0.078125
Errors at time T: L1-error = 0.038395, max. error = 0.1054

etaIOS, wave, dx=0.125, dy=0.125, dt=0.078125
Errors at time T: L1-error = 0.011013, max. error = 0.027888
```

```
etaIOS, wave, dx=0.0625, dy=0.0625, dt=0.078125
Errors at time T: L1-error = 0.0038664, max. error = 0.0096016
```

$L_A T_E X$-code for the table and the plots is also possible.

TABLE 8.1:   Numerical results for the wave equation with coefficients $(D_1, D_2) = (1, 1)$ using the etaIOS method, initial condition $U_{n+1,0}(t) = U_n$, and two iterations per time step.

$\Delta x = \Delta y$	Δt	err_{L_1}	$\text{err}_{\max}$	ρ_{L_1}	$\rho_{\max}$
0.25	0.3125	0.048967	0.13442		
0.125	0.3125	0.021581	0.054647	1.1821	1.2985
0.0625	0.3125	0.014382	0.035715	0.58548	0.61361
0.25	0.15625	0.042097	0.11556		
0.125	0.15625	0.014849	0.037601	1.5034	1.6199
0.0625	0.15625	0.0077193	0.01917	0.94382	0.97195
0.25	0.078125	0.038395	0.1054		
0.125	0.078125	0.011013	0.027888	1.8017	1.9182
0.0625	0.078125	0.0038664	0.0096016	1.5102	1.5383

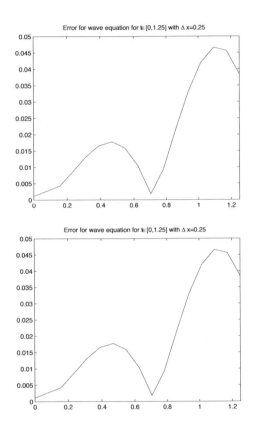

FIGURE 8.3: Two of the figures, which are automatically created by FIDOS.

Appendix

List of Abbreviations

- ADI: Alternating direction implicit methods

- CFL condition: Courant-Friedrichs-Levy condition

- COM: Component object model

- CVD: Chemical vapor deposition

- DD: Domain decomposition methods

- D^3F: Distributed-density-driven flow software toolbox, done with the software package UG (unstructured grids)

- DLL: Dynamic link library

- GRAPE: Visualization software, see [116]

- ILU: Incomplete LU method

- IOS: Iterative operator splitting method

- LOD: Locally one-dimensional methods, e.g., locally one-dimensional splitting methods

- MAX-phase: Special material with metallic and ceramic behavior, see [12]

- MD: Molecular dynamics

- ODE: Ordinary differential equation

- OFELI: Object finite element library

- OPERA-SPLITT: Software package including splitting methods

- PDE: Partial differential equation

- PVD: Physical vapor deposition

- R^3T: Radioactive-reaction-retardation-transport software toolbox, done with the software package UG (unstructured grids)

- SiC: Silicon carbide

- Ti_3SiC_2: Special material used for thin-layer deposition, see [12]

Symbols

- A - in the following A is a matrix in $\mathbb{R}^m \times \mathbb{R}^m$, $m \in \mathbb{N}^+$ is the rank

- λ_i - i-th eigenvalue of A

- $\rho(A)$ - spectral radius of A

- e_i - i-th eigenvector of matrix A

- $\sigma(A)$ - spectrum of A

- $Re(\lambda_i)$ - i-th real eigenvalue of λ

- $u_t = \frac{\partial u}{\partial t}$ - first-order partial time derivative of c

- $u_{tt} = \frac{\partial^2 u}{\partial t^2}$ - second-order partial time derivative of c

- $u_{ttt} = \frac{\partial^3 u}{\partial t^3}$ - third-order partial time derivative of u

- $u_{tttt} = \frac{\partial^4 u}{\partial t^4}$ - fourth-order partial time derivative of u

- $u' = \frac{du}{dt}$ - first-order time derivative of u

- $u'' = \frac{d^2 u}{dt^2}$ - second-order time derivative of u

- $\tau = \tau_n = t^{n+1} - t^n$ - time step

- u^n - approximated solution of u at time t^n

- $\partial_t^+ u = \frac{u^{n+1} - u^n}{\tau_n}$ - forward finite difference of u in time

- $\partial_t^- u = \frac{u^n - u^{n-1}}{\tau_n}$ - backward finite difference of u in time

- $\partial_t^0 u = \frac{u^{n+1} - u^{n-1}}{2\tau_n}$ - central finite difference of u in time

- $\partial_t^2 u = \partial_t^+ \partial_t^- u$ - second-order finite difference of u in time

- ∇u - gradient of u

- $\Delta u(x, t)$ - Laplace operator of u

- $\nabla \cdot \mathbf{u}$ - divergence of $\mathbf{u}$ (where $\mathbf{u}$ is a vector function)

- n_m - outer normal vector to Ω_m

- $\partial_x^+ u$ - forward finite difference of u in space dimension x

- $\partial_x^- u$ - backward finite difference of u in space dimension x

- $\partial_x^0 u$ - central finite difference of u in space dimension x

- $\partial_x^2 u$ - second-order finite difference of u in space dimension x

- $\partial_y^+ u$ - forward finite difference of u in space dimension y

- $\partial_y^- u$ - backward finite difference of u in space dimension y

- $\partial_y^0 u$ - central finite difference of u in space dimension y

- $\partial_y^2 u$ - second-order finite difference of u in space dimension y

- $e_i(t) := u(t) - u_i(t)$ - local error function with approximated solution $u_i(t)$

- err_{local} - local splitting error

- err_{global} - global splitting error

- $arg(z)$ - argument of $z \in \mathbb{C}$ (angle between it and the real axis)

- $[A, B] = AB - BA$ - commutator of operator A and B

General Notations

$D(B)$	Domain of B
$\mathbf{X}, \mathbf{X}_E$	Banach spaces
$\mathbf{X}^n = \Pi_{i=1}^n \mathbf{X}_i$	Product space of $\mathbf{X}$
$W^{m,p}(\Omega)$	Sobolev space consisting of all locally summable functions $u : \Omega \to \mathbb{R}$ such that for each multi-index α with $|\alpha| \le m$, $\partial_\alpha u$ exists in the weak sense and belongs to $L^p(\Omega)$
$\partial\Omega$	Boundary of Ω
$\mathcal{L}(\mathbf{X}) = L(\mathbf{X}, \mathbf{X})$	Operator space of $\mathbf{X}$, e.g., a Banach space
Ω_h	Discretized domain Ω with the underlying grid step h
H^m	Sobolev space $W^{m,2}$
$H_0^1(\Omega)$	The closure of $C_c^\infty(\Omega)$ in the Sobolev space $W^{1,2}$.
$\|\cdot\|_{L^p}$	L^p-norm
$\|\cdot\|_{H^m}$	H^m-norm
$\|\cdot\|$	Maximum norm, if not defined otherwise
$\|\cdot\|_{\mathbf{X}}$	Norm with respect to Banach space $\mathbf{X}$
$\|\cdot\|_\infty = \sup_{t \in I} \|\cdot\|$	Maximum norm on interval I
(x, y)	Scalar product of x and y in a Hilbert space, see Chapter 2
$\mathcal{O}(\tau)$	Landau symbol, e.g., first order in time with time step τ
$U = (u, v)^T$	Vectorial solutions with two components, see Chapter 2
$U = (u, v, w)^T$	Vectorial solutions in three components, see Chapter 2
$(x_1, \ldots, x_n)^T = \begin{pmatrix} x_1 \\ \vdots \\ x_n \end{pmatrix}$	Vectorial notation, see Chapter 2 and 6

Bibliography

[1] M. Abramowitz and I.A. Stegun. *Handbook of Mathematical Functions.* Dover Publication, New York, 1970.

[2] N.I. Akhiezer and I.M. Glazman. *Theory of Linear Operators in Hilbert Space.* Dover Publications, 1993.

[3] M. Allen and D. Tildesly. *Computer Simulation of Liquids.* Clarendon Press, Oxford, 1987.

[4] I. Alonso-Mallo, B. Cano, and J.C. Jorge. *Spectral-fractional step Runge-Kutta discretisations for initial boundary value problems with time dependent boundary conditions.* Mathematics of Computation, 73, 1801–1825, 2004.

[5] Z. S. Alterman and A. Rotenberg. *Seismic waves in a quarter plane.* Bulletin of the Seismological Society of America, 59, 347–368, 1969.

[6] N. Antonic, C.J. van Duijn, W. Jäger, and A. Mikelic. *Multiscale Problems in Science and Technology: Challenges to Mathematical Analysis and Perspectives.* Proceedings of the Conference on Multiscale Problems in Science and Technology, Dubrovnik, Croatia, 3-9 September 2000, N. Antonic, C.J. van Duijn, W. Jäger and A. Mikelic (Eds.), Springer-Verlag Berlin-Heidelberg-New York, 2002.

[7] U.M. Ascher, S.J. Ruuth, and R.J. Spiteri. *Implicit-explicit Runge-Kutta methods for time-dependent partial differential equations.* Applied Numerical Mathematics, 25, 151–167, 1997.

[8] O. Axelsson. *Iterative Solution Methods.* Cambridge University Press, 1996.

[9] W. Balser, and J. Mozo-Fernandez. *Multisummability of formal solutions of singular perturbation problems.* J. Differential Equations, 183(2), 526–545, 2002.

[10] W. Balser, A. Duval, and St. Malek. *Summability of formal solutions for abstract Cauchy problems and related convolution equations.* Manuscript, November 2006.

[11] V. Barbu. *Nonlinear Semigroups and Differential Equations in Banach Spaces.* Editura Academiei, Bucuresti, and Noordhoff, Leyden, 1976.

[12] M.W. Barsoum and T. El-Raghy. *Synthesis and characterization of a remarkable ceramic: Ti_3SiC_2.* J. Am. Ceram. Soc., 79(1), 1953–1956, 1996.

[13] D.A. Barry, C.T. Miller, and P.J. Culligan-Hensley. *Temporal discretization errors in non-iterative split-operator approaches to solving chemical reaction/groundwater transport models.* Journal of Contaminant Hydrology, 22, 1–17, 1996.

[14] P. Bastian. *Parallele adaptive Mehrgitterverfahren.* Doktor-Arbeit, Universität Heidelberg, 1994.

[15] P. Bastian, K. Birken, K. Eckstein, K. Johannsen, S. Lang, N. Neuss, and H. Rentz-Reichert. *UG - a flexible software toolbox for solving partial differential equations.* Computing and Visualization in Science, 1(1), 27–40, 1997.

[16] R.M. Beam and R. F. Warming. *Alternating direction implicit methods for parabolic equations with a mixed derivative.* SIAM J. Sci. Stat.Comput., 1, 131–159, 1980.

[17] J. Bear. *Dynamics of Fluids in Porous Media.* American Elsevier, New York, 1972.

[18] J. Bear and Y. Bachmat. *Introduction to Modeling of Transport Phenomena in Porous Media.* Kluwer Academic Publishers, Dordrecht, Boston, London, 1991.

[19] H. Berendsen, J. Postma, W. van Gunsteren, A. Dinola, and J. Haak. *Molecular-dynamics with coupling to an external bath.* J. Chem. Phys., 81, 3684–3690, 1984.

[20] M. Bjorhus. *Operator splitting for abstract Cauchy problems.* IMA Journal of Numerical Analysis, 18, 419–443, 1998.

[21] D. Buhmann. *Das Programmpaket EMOS. Ein Instrumentarium zur Analyse der Langzeitsicherheit von Endlagern.* Gesellschaft für Anlagen- und Reaktorsicherheit (mbH), GRS-159, Braunschweig, 1999.

[22] J.C. Butcher. *Implicit Runge-Kutta processes.* Math. Comp., 18, 50–64, 1964.

[23] J.C. Butcher. *Numerical Methods for Ordinary Differential Equations.* John Wiley & Sons Ltd, Chichester, 2003.

[24] C.X. Cao. *A theorem on the separation of a system of coupled differential equations.* J. Phys. A.: Math. Gen., 14, 1069–1074, 1981.

[25] G.Z. Cao, H. Brinkman, J. Meijerink, K.J. DeVries, and A.J. Burggraaf. *Kinetic study of the modified chemical vapour deposition process in porous media*. J.Mater.Chem., 3(12), 1307–1311, 1993.

[26] M.A. Celia, J.S. Kindred, and I. Herrera. *Contaminant transport and biodegradation, 1. A numerical model for reactive transport in porous media*. Water Resources Research, 25, 1141–1148, 1989.

[27] Q.-S. Chen, H. Zhang, V. Prasad, C.M. Balkas, and N.K. Yushin. *Modeling of heat transfer and kinetics of physical vapor transport growth of silicon carbide crystals*. Transactions of the ASME. Journal of Heat Transfer, vol. 123, No. 6, 1098–1109, 2001.

[28] S.A. Chin. *A fundamental theorem on the structure of symplectic integrators*. Physics Letters A, 354, 373–376, 2006.

[29] S.A. Chin and C.R. Chen. *Gradient symplectic algorithms for solving the Schrödinger equation with time-dependent potentials*. The Journal of Chemical Physics, 117(4), 1409–1415, 2002.

[30] S. Chin and J. Geiser. *Multi-product operator splitting as a general method of solving autonomous and non-autonomous equations*. IMA J. Numer. Anal., first published online January 12, 2011.

[31] W. Cheney. *Analysis for Applied Mathematics*. Graduate Texts in Mathematics., 208, Springer, New York, Berlin, Heidelberg, 2001.

[32] K.H. Coats and B.D. Smith. *Dead-end pore volume and dispersion in porous media*. Society of Petroleum Engineers Journal, 4(3), 73–84, 1964.

[33] G.C. Cohen. *Higher-Order Numerical Methods for Transient Wave Equations*. Springer-Verlag, 2002.

[34] R. Courant, K.O. Friedrichs, and H. Lewy. Collatz. *Über die partiellen Differenzengleichungen der mathematischen Physik*. Math.Ann., 100, 32–74, 1928.

[35] D. Daoud and J. Geiser. *Overlapping schwarz wave form relaxation for the solution of coupled and decoupled system of convection diffusion reaction equation*. Applied Mathematics and Computation, Elsevier, North Holland, 190(1), 946–964, 2007.

[36] B. Davis. *Integral Transform and Their Applications*. Applied Mathematical Sciences, Springer-Verlag, New York, Heidelberg, Berlin, No. 25, 1978.

[37] St.M. Day, et al. *Test of 3D elastodynamic codes: Final report for lifelines project 1A01*. Technical report, Pacific Earthquake Engineering Center, 2001.

[38] St.M. Day, et al. *Test of 3D elastodynamic codes: Final report for lifelines project 1A02*. Technical report, Pacific Earthquake Engineering Center, 2003.

[39] K. Dekker and J.G. Verwer. *Stability of Runge-Kutta Methods for Stiff Nonlinear Differential Equations*. North-Holland Elsevier Science Publishers, Amsterdam, New York, Oxford, 1984.

[40] J. Douglas, Jr. and S. Kim. *Improved accuracy for locally one-dimensional methods for parabolic equations*. Mathematical Models and Methods in Applied Sciences, 11, 1563–1579, 2001.

[41] F. Dupret, P. Nicodéme, Y. Ryckmans, P. Wouters, and M.J. Crochet. *Global modelling of heat transfer in crystal growth furnaces*. Intern. J. Heat Mass Transfer, 33(9), 1849–1871, 1990.

[42] M.K. Dobkin and D.M. Zuraw. *Principles of Chemical Vapor Deposition*. Springer-Verlag, Heidelberg, New York, First Edition, 2003.

[43] J.R. Dormand and P.J. Prince. *A family of embedded Runge-Kutta formulae*. Journal of Computational and Applied Mathematics, 6(1), 19-26, 1980.

[44] D.R. Durran. *Numerical Methods for Wave Equations in Geophysical Fluid Dynamics*. Springer-Verlag, New-York, Heidelberg, 1998.

[45] E. G. D'Yakonov. *Difference schemes with splitting operator for multidimensional nonstationary problems*. Zh. Vychisl. Mat. i. Mat. Fiz., 2, 549–568, 1962.

[46] G. Eason, J. Fulton, and I. N. Sneddon. *The Generation of waves in an infinite elastic solid by variable body forces*. Phil. Trans. R. Soc. Lond., 248(955), 575–607, 1956.

[47] G.R. Eykolt. *Analytical solution for networks of irreversible first-order reactions*. Wat. Res., 33(3), 814–826, 1999.

[48] G.R. Eykolt and L. Li. *Fate and transport of species in a linear reaction network with different retardation coefficients*. Journal of Contaminant Hydrology, 46, 163–185, 2000.

[49] K.-J. Engel and R. Nagel, *One-Parameter Semigroups for Linear Evolution Equations*. Springer, New York, 2000.

[50] L.C. Evans. *Partial Differential Equations*. Graduate Studies in Mathematics, Volume 19, AMS, 1998.

[51] R.E. Ewing. *Up-scaling of biological processes and multiphase flow in porous media*. *IIMA Volumes in Mathematics and its Applications*, Springer-Verlag, 295, 195–215, 2002.

[52] G. Fairweather and A.R. Mitchell. *A high accuracy alternating direction method for the wave equations.* J. Industr. Math. Appl., 1, 309–316, 1965.

[53] I. Farago. *Splitting methods for abstract Cauchy problems.* Lect. Notes Comp. Sci., Springer-Verlag, Berlin, 3401, 35-45, 2005.

[54] I. Farago. *Modified iterated operator splitting method.* Applied Mathematical Modelling, 32(8), 1542–1551, 2008.

[55] I. Farago and J. Geiser. *Iterative Operator-splitting methods for linear problems.* International Journal of Computational Science and Engineering, 3(4), 255–263, 2007.

[56] I. Farago and A. Havasi. *Consistency analysis of operator splitting methods for C_0-semigroups.* Semigroup Forum, 74, 125–139, 2007.

[57] R. Fazio and A. Jannelli. *Second order positive schemes by means of flux limiters for the advection equation.* IANG International Journal of Applied Mathematics, 39(1), 25–35, 2009.

[58] E. Fein and A. Schneider. *d^3f - Ein Programmpaket zur Modellierung von Dichteströmungen.* Abschlussbericht, Braunschweig, 1999.

[59] E. Fein, T. Kühle, and U. Noseck. *Entwicklung eines Programms zur dreidimensionalen Modellierung des Schadstofftransportes.* Fachliches Feinkonzept, Braunschweig, 2001.

[60] E. Fein. *Beispieldaten für radioaktiven Zerfall.* Private communications, Braunschweig, 2000.

[61] E. Fein. *Physikalisches Modell und mathematische Beschreibung.* Private communications, Braunschweig, 2001.

[62] E. Fein. *Software Package r^3t: Model for Transport and Retention in Porous Media.* Final Report, GRS-192, Braunschweig, 2004.

[63] P. Frolkovič and J. Geiser. *Numerical Simulation of Radionuclides Transport in Double Porosity Media with Sorption.* Proceedings of Algorithmy 2000, Conference of Scientific Computing, 28–36, 2000.

[64] M.J. Gander and A.M. Stuart. *Space-time continuous analysis of waveform relaxation for the heat equation.* SIAM Journal on Scientific Computing, 19(6), 2014–2031, 1998.

[65] M.J. Gander and S. Vanderwalle. *Analysis of the parareal time-parallel time-integration method.* SIAM Journal of Scientific Computing, 29(2), 556–578, 2007.

[66] M.J. Gander and H. Zhao. *Overlapping Schwarz waveform relaxation for parabolic problems in higher dimension.* In A. Handlovičová, Magda Komorníkova, and Karol Mikula, editors, in: Proc. Algoritmy 14, Slovak Technical University, 42–51, 1997.

[67] J. Geiser. *Numerical Simulation of a Model for Transport and Reaction of Radionuclides.* Lecture Notes in Computer Science, vol. 2179, Proceedings of the Third International Conference on Large-Scale Scientific Computing, Sozopol, Bulgaria, 487–496, 2001.

[68] J. Geiser. *Gekoppelte Diskretisierungsverfahren für Systeme von Konvektions-Dispersions-Diffusions-Reaktionsgleichungen.* PhD thesis, University of Heidelberg, Germany, 2004.

[69] J. Geiser. R^3T: *Radioactive-Retardation-Reaction-Transport-Program for the Simulation of radioactive waste disposals.* Technical report, ISC-04-03-MATH, Institute for Scientific Computation, Texas A&M University, College Station, TX, 2004.

[70] J. Geiser. *Discretisation Methods with Embedded Analytical Solutions for Convection Dominated Transport in Porous Media.* Lect. Notes in Mathematics, Springer, vol. 3401, 288–295, 2005, Proceedings of the 3rd International Conference, NAA 2004, Rousse, Bulgaria.

[71] J. Geiser, R.E. Ewing, and J. Liu. *Operator Splitting Methods for Transport Equations with Nonlinear Reactions.* Proceedings of the Third MIT Conference on Computational Fluid and Solid Mechanic, Cambridge, MA, June 14–17, 2005.

[72] J. Geiser and J. Gedicke. *Nonlinear Iterative Operator-Splitting Methods and Applications for Nonlinear Parabolic Partial Differential Equations.* Preprint No. 2006-17 of Humboldt University of Berlin, Department of Mathematics, Germany.

[73] J. Geiser. *Discretization methods with analytical solutions for convection-diffusion-dispersion-reaction-equations and application.* Journal of Engineering Mathematics, 57, 79–98, 2007.

[74] J. Geiser. *Weighted Iterative Operator-Splitting Methods: Stability-Theory.* Lecture Notes in Computer Science (Springer), vol. 4310, 40–47, 2007, Proceedings of the 6th International Conference, NMA 2006, Borovets, Bulgaria.

[75] J. Geiser and Chr. Kravvaritis. *Weighted Iterative Operator-Splitting Methods and Applications.* Lecture Notes in Computer Science (Springer), vol. 4310, 48–55, 2007, Proceedings of the 6th International Conference, NMA 2006, Borovets, Bulgaria.

[76] J. Geiser and St. Nilsson. *A Fourth Order Split Scheme for Elastic Wave Propagation.* Preprint 2007-08, Humboldt University of Berlin, Department of Mathematics, Germany, 2007.

[77] J. Geiser and L. Noack. *Iterative operator-splitting methods for wave equations with stability results and numerical examples.* Preprint 2007-10, Humboldt-University of Berlin, 2007.

[78] J. Geiser and V. Schlosshauer. *Operator-Splitting Methods for Wave-Equations.* Preprint 2007-06, Humboldt University of Berlin, Department of Mathematics, Germany, 2007.

[79] J. Geiser and S. Sun. *Multiscale Discontinuous Galerkin Methods for Modeling Flow and Transport in Porous Media.* Lecture Notes in Computational Science, 4487, 890–897, 2007.

[80] J. Geiser. *Operator splitting methods for wave equations.* International Mathematical Forum, Hikari Ltd., 2(43), 2141–2160, 2007.

[81] J. Geiser. *Weighted Iterative Operator-Splitting Methods: Stability-Theory.* Lecture Notes in Computer Science, Springer Verlag, 4310, 40–47, 2007, Proceedings of the 6th International Conference, NMA 2006, Borovets, Bulgaria.

[82] J. Geiser. *Iterative operator-splitting methods with higher order time-integration methods and applications for parabolic partial differential equations.* Journal of Computational and Applied Mathematics, Elsevier, Amsterdam, The Netherlands, 217, 227–242, 2008.

[83] J. Geiser and L. Noack. *Iterative Operator-Splitting Methods for Nonlinear Differential Equations and Applications of Deposition Processes.* Preprint 2008-04, Humboldt-University of Berlin, 2008.

[84] J. Geiser and L. Noack. *Operator-Splitting Methods Respecting Eigenvalue Problems for Nonlinear Equations and Application in Burgers-Equations.* Preprint 2008-13, Humboldt-University of Berlin, 2008.

[85] J. Geiser. *Fourth-order splitting methods for time-dependent differential equations.* Numerical Mathematics: Theory, Methods and Applications, 1(3), 321–339, 2008.

[86] J. Geiser. *Iterative operator-splitting methods with higher order time-integration methods and applications for parabolic partial differential equations.* Journal of Computational and Applied Mathematics, Elsevier, Amsterdam, The Netherlands, 217, 227–242, 2008.

[87] J. Geiser and Chr. Kravvaritis. *Overlapping operator splitting methods and applications in stiff differential equations.* Special issue: Novel Difference and Hyprod Methods for Differential and Integro-Differential Equations and Applications, Guest editors: Qin Sheng and Johnny Henderson, Neural, Parallel, and Scientific Computations (NPSC), 16, 189–200, 2008.

[88] J. Geiser. *Discretization and Simulation of Systems for Convection-Diffusion-Dispersion Reactions with Applications in Groundwater Contamination.* Series: Groundwater Modelling, Management and Contamination, Nova Science Publishers, Inc. New York, Monograph, 2008.

[89] J. Geiser. *Stability of Iterative Operator-Splitting Methods.* International Journal of Computer Mathematics, 1029-0265, First published on 26 June 2009, http://www.informaworld.com, 2009.

[90] J. Geiser. *Computation of Iterative Operator-Splitting Methods.* Preprint 2009-21, Humboldt University of Berlin, Department of Mathematics, Germany, 2009.

[91] J. Geiser and Chr. Kravvaritis. *A domain decomposition method based on iterative operator splitting method.* Applied Numerical Mathematics, 59, 608–623, 2009.

[92] J. Geiser. *Iterative Operator-Splitting with Time Overlapping Algorithms: Theory and Application to Constant and Time-Dependent Wave Equations.* Wave Propagation in Materials for Modern Applications, Andrey Petrin (Ed.), ISBN: 978-953-7619-65-7, INTECH, 2009.

[93] J. Geiser. *Operator-splitting methods in respect of eigenvalue problems for nonlinear equations and applications to Burgers equations.* Journal of Computational and Applied Mathematics, Elsevier, Amsterdam, North Holland, 231(2), 815–827, 2009.

[94] J. Geiser and R. Steijl. *Coupled Navier Stokes - Molecular Dynamics Simulation using Iterative Operator-Splitting Methods.* Preprint 2009-11, Humboldt University of Berlin, Department of Mathematics, Germany, 2009.

[95] J. Geiser. *Decomposition Methods for Partial Differential Equations: Theory and Applications in Multiphysics Problems.* Numerical Analysis and Scientific Computing Series, CRC Press, Chapman & Hall/CRC, edited by Magoules and Lai, 2009.

[96] J. Geiser. *Discretization methods with analytical characteristic methods and applications.* M2AN, EDP Sciences, France, 43(6), 1157–1183, 2009.

[97] J. Geiser and S. Chin. *Multi-Product Expansion, Suzuki's Method and the Magnus Integrator for Solving Time-Dependent Problems.* Preprint 2009-4, Humboldt University of Berlin, Department of Mathematics, Germany, 2009.

[98] J. Geiser and G. Tanoglu. *Iterative Operator-Splitting Methods, Continuous and Discrete Case: Theory and Applications.* Preprint 2009-13, Humboldt University of Berlin, Department of Mathematics, Germany, 2009.

[99] J. Geiser and G. Tanoglu. *Higher Order Operator-Splitting Methods via Zassenhaus Product Formula: Theory and Applications.* Preprint 2009-15, Humboldt University of Berlin, Department of Mathematics, Germany, 2009.

[100] J. Geiser and G. Tanoglu. *Successive Approximation for Solving Time-Dependent Problems: Theoretical Overview.* Proceeding, Fifth Conference on Finite Difference Methods: Theory and Applications, June 28-July 2, 2010, Lozenetz, Bulgaria, accepted November 2010.

[101] J. Geiser and M. Arab. *Modelling, optimization and simulation for a chemical vapor deposition.* Journal of Porous Media, Begell House Inc., Redding, USA, 12(9), 847–867, 2009.

[102] J. Geiser and F. Krien. *Iterative Operator-Splitting Methods for Time-Irreversible Systems: Theory and Application to Advection-Diffusion Equations.* Preprint 2009-18, Humboldt University of Berlin, Department of Mathematics, Germany, 2009.

[103] J. Geiser and F. Röhle. *Kinetic Processes and Phase-Transition of CVD Processes for Ti_3SiC_2.* Preprint 2010-1, Humboldt University of Berlin, Department of Mathematics, Germany, 2010.

[104] J. Geiser. *Models and Simulation of Deposition Processes with CVD Apparatus.* Monograph, Series: Groundwater Modelling, Management and Contamination, Nova Science Publishers, New York, 2009.

[105] J. Geiser. *Iterative Operator-Splitting Methods for Nonlinear Differential Equations and Applications.* Numerical Methods for Partial Differential Equations, John Wiley & Sons, Ltd., West Sussex, UK, published online, March 2010.

[106] J. Geiser. *Mobile and immobile fluid transport: Coupling framework.* International Journal for Numerical Methods in Fluids, accepted as Review October 2009, online published (http://www3.interscience.wiley.com/cgi-bin/fulltext/123276563/PDFSTART), 2010.

[107] J. Geiser. *Consistency of iterative operator-splitting methods: Theory and applications.* Numerical Methods for Partial Differential Equations, 26(1), 135–158, 2010.

[108] J. Geiser and M. Arab. *Simulation of a chemical vapor deposition: mobile and immobile zones and homogeneous layers.* Special Topics and Reviews in Porous Media, Begell House Inc., Redding, USA, 1(2), 123–143, 2010.

[109] J. Geiser and M. Arab. *Porous media based modeling of PE-CVD apparatus: electrical fields and deposition geometries.* Special Topics and Reviews in Porous Media, Begell House Inc., Redding, USA, 1(3), 215–229, 2010.

[110] J. Geiser. *Magnus Integrator and Successive Approximation for Solving Time-Dependent Problems.* Preprint 2010-10, Humboldt University of Berlin, Department of Mathematics, Germany, 2010.

[111] J. Geiser. *Decomposition Methods in Multiphysics and Multiscale Problems.* Series: Physics Research and Technology, Nova Science Publishers, Inc. New York, Monograph, 2010.

[112] J. Geiser. *Computing Exponential for Iterative Splitting Methods.* Journal of Applied Mathematics, special issue: Mathematical and Numerical Modeling of Flow and Transport (MNMFT), Hindawi Publishing Corp., New York, accepted, January 2011.

[113] P. George. *Chemical Vapor Deposition: Simulation and Optimization.* VDM Verlag Dr. Müller, Saarbrücken, Germany, First Edition, 2008.

[114] R. Glowinski. *Numerical Methods for Fluids.* Handbook of Numerical Analysis, Gen. eds. P.G. Ciarlet, J. Lions, Vol. IX, North-Holland Elsevier, Amsterdam, The Netherlands, 2003.

[115] M.K. Gobbert and C.A. Ringhofer. *An asymptotic analysis for a model of chemical vapor deposition on a microstructured surface.* SIAM Journal on Applied Mathematics, 58, 737–752, 1998.

[116] GRAPE. *GRAphics Programming Environment for mathematical problems, Version 5.4.* Institut für Angewandte Mathematik, Universität Bonn und Institut für Angewandte Mathematik, Universität Freiburg, 2001.

[117] Chr. Grossmann, H.G. Roos, and M. Stynes. *Numerical Treatment of Partial Differential Equations.* Universitext, first edition, Springer-Verlag Berlin-Heidelberg-New York, 2007.

[118] B. Gustafsson. *High Order Difference Methods for Time Dependent PDE.* Springer Series in Computational Mathematics, Springer-Verlag, Berlin, New York, Heidelberg, 38, 2007.

[119] W. Hackbusch. *Multi-Grid Methods and Applications.* Springer-Verlag, Berlin, Heidelberg, 1985.

[120] E. Hansen and A. Ostermann. *Exponential splitting for unbounded operators.* Math. Comp. 78, 1485–1496, 2009.

[121] E. Hansen and A. Ostermann. *High order splitting methods for analytic semigroups exist.* BIT, 49, 527–542, 2009.

[122] E. Hairer, S.P. Norsett, and G. Wanner. *Solving Ordinary Differential Equations I*. SCM, Springer-Verlag Berlin-Heidelberg-New York, No. 8, 1992.

[123] E. Hairer and G. Wanner. *Solving Ordinary Differential Equations II*. SCM, Springer-Verlag Berlin-Heidelberg-New York, No. 14, 1996.

[124] E. Hairer, C. Lubich, and G. Wanner. *Geometric Numerical Integration: Structure-Preserving Algorithms for Ordinary Differential Equations*. SCM, Springer-Verlag Berlin-Heidelberg-New York, No. 31, 2002.

[125] A. Harten. *High resolution schemes for hyperbolic conservation laws*. Journal of Computational Physics, 135(2), 260–278, 1997.

[126] A. Havasi, J. Bartholy, and I. Farago. *Splitting method and its application in air pollution modeling*. Quarterly Journal of the Hungarian Meteorological Service, 105(1), 39–58, January–March 2001.

[127] D. Henry. *Geometric Theory of Semilinear Parabolic Equations*, Lecture Notes in Mathematics, Springer-Verlag, Berlin, 1981.

[128] J. Herzer and W. Kinzelbach. *Coupling of transport and chemical processes in numerical transport models*. Geoderma, 44, 115–127, 1989.

[129] V. Hlavacek, J. Thiart, and D. Orlicki. *Morphology and Film Growth in CVD Reactions*. J. Phys. IV France, 5, 3–44, 1995.

[130] M. Hochbruck and C. Lubich. *Exponential integrators for large systems of differential equations*. SIAM J. Sci. Comput., 19(5), 1552–1574, 1998.

[131] M. Hochbruck and C. Lubich. *On Krylov subspace approximations to the matrix exponential operator*. SIAM J. Numer. Anal., 34(5), 1911–1925, 1997.

[132] M. Hochbruck and A. Ostermann. *Explicit Exponential Runge-Kutta Methods for Semilinear Parabolic Problems*. SIAM Journal on Numerical Analysis, 43(3), 1069–1090, 2005.

[133] W. Hundsdorfer and J.G. Verwer. *Numerical Solution of Time-Dependent Advection-Diffusion-Reaction Equations*. Springer-Verlag, Berlin, 2003.

[134] W. Hundsdorfer and L. Portero. *A note on iterated splitting schemes*. Journal of Computational and Applied Mathematics, 201(1), 146–152, 2007.

[135] J. Irving and J. Kirkwood. *The statistical mechanical theory of transport processes IV*. J. Chem. Phys., 18, 817–829, 1950.

[136] T. Jahnke and C. Lubich. *Error bounds for exponential operator splittings*. BIT Numerical Mathematics, 40(4), 735–745, 2000.

[137] H.A. Jakobsen. *Chemical Reactor Modeling: Multiphase Reactive Flows.* Springer-Verlag, Heidelberg, New York, 1st edition, 2008.

[138] A. Jameson. *Time dependent calculations using multigrid with application to unsteady flows past airfoils and wings.* AIAA Paper 91-1596, 1991.

[139] J. Janssen and S. Vandewalle. *Multigrid waveform relaxation on spatial finite-element meshes: The continuous case.* SIAM J. Numer. Anal., 33, 456–474, 1996.

[140] Y.L. Jiang. *Periodic waveform relaxation solutions of nonlinear dynamic equations.* Applied Mathematics and Computation, 135(2–3), 219–226, 2003.

[141] K. Johannsen. *Robuste Mehrgitterverfahren für die Konvektions-Diffusions Gleichung mit wirbelbehafteter Konvektion.* PhD thesis, University of Heidelberg, Germany, 1999.

[142] K. Johannsen. *An aligned 3D-finite-volume method for convection-diffusion problems.* Modeling and Computation in Environmental Sciences, R. Helmig, W. Jäger, W. Kinzelbach, P. Knabner, G. Wittum (eds.), Vieweg, Braunschweig, 59, 227–243, 1997.

[143] S.L. Johnson, Y. Saad, and M. Schultz. *Alternating direction methods on multiprocessors.* SIAM J. Sci. Stat. Comput., 8(5), 686–700, 1987.

[144] J. Kanney, C. Miller, and C.T. Kelley. *Convergence of iterative split-operator approaches for approximating nonlinear reactive transport problems.* Advances in Water Resources, 26, 247–261, 2003.

[145] K.H. Karlsen, K.-A. Lie, J.R. Natvig, H.F. Nordhaug, and H.K. Dahle. *Operator splitting methods for systems of convection-diffusion equations: nonlinear error mechanisms and correction strategies.* Journal of Computational Physics, 173(2), 636–663, 2001.

[146] C.T. Kelley. *Iterative Methods for Linear and Nonlinear Equations.* SIAM Frontiers in Applied Mathematics, no. 16, SIAM, Philadelphia, 1995.

[147] C.T. Kelley. *Solving Nonlinear Equations with Newton's Method.* Computational Mathematics, SIAM, XIV, 2003.

[148] S. Kim and H. Lim. *High-order schemes for acoustic waveform simulation.* Applied Numerical Mathematics, 57(4), 402–414, 2007.

[149] J. Herzer and W. Kinzelbach. *Coupling of transport and chemical processes in numerical transport models.* Geoderma, 44, 115–127, 1989.

[150] W. Kinzelbach. *Numerische Methoden zur Modellierung des Transports von Schadstoffen im Grundwasser.* Schriftenreihe Wasser-Abwasser, Oldenburg, 1992.

[151] R. Kozlov and B. Owren. *Order Reduction in Operator Splitting Methods.* Preprint N6-1999, Department of Mathematical Sciences, Norwegian University of Science and Technology, Trondheim, Norway, 1999.

[152] R. Kozlov, A. Kvarno, and B. Owren. *The behaviour of the local error in splitting methods applied to stiff problems.* Journal of Computational Physics, 195, 576–593, 2004.

[153] D. Lanser and J.G. Verwer. *Analysis of operator splitting for advection-diffusion-reaction problems from air pollution modelling.* Journal of Computational Applied Mathematics, 111(1–2), 201–216, 1999.

[154] L. Lapidus and G.F. Pinder. *Numerical Solution of Partial Differential Equations in Science and Engineering.* John Wiley & Sons, Incorporation, USA, 1996.

[155] M. Lees. *Alternating direction methods for hyperbolic differential equations.* J. Soc. Industr. Appl. Math., 10(4), 610–616, 1962.

[156] P. van Leemput, W. Vanroose and D. Roose. *Mesoscale analysis of the equation-free constrained runs initialization scheme* Multiscale Model. Simul. MODEL., 6(4), 1234–1255, 2008.

[157] A. Lees and S. Edwards. *The computer study of transport processes under extreme conditions.* J. Phys. C., 5, 1921–1929, 1972.

[158] E. Lelarasmee, A. Ruehli, and A. Sangiovanni-Vincentelli. *The waveform relaxation methods for time domain analysis of large scale integrated circuits.* IEEE Trans. CAD IC Syst., 1, 131–145, 1982.

[159] J. Lennard-Jones. *Cohesion.* Proc. Phys. Soc., 43, 461–482, 1931.

[160] R.J. LeVeque. *Finite Volume Methods for Hyperbolic Problems.* Cambridge Texts in Applied Mathematics, Cambridge University Press, 2002.

[161] R.J. LeVeque. *Finite Difference Methods for Ordinary and Partial Differential Equations, Steady State and Time Dependent Problems.* Society for Industrial and Applied Mathematics (SIAM), Philadelphia, 2007.

[162] M.A. Lieberman and A.J. Lichtenberg. *Principle of Plasma Discharges and Materials Processing.* Wiley-Interscience, John Wiley & Sons, Inc. Publication, second edition, 2005.

[163] P. Lindelöf. *Sur l'application des methodes d'approximations successives a l'etude de certaines equations differentielles ordinaires.* J. de Math. Pures et Appl., 4(9), 217–271, 1893.

[164] P. Lindelöf. *Sur l'application des methodes d'approximations successives a l'etude des integrales reelles des equations differentielles ordinaires.* J. de Math. Pures et Appl., 4(10), 117–128, 1894.

[165] W. Magnus. *On the exponential solution of differential equations for linear operator.* Communication on pure and applied mathematics, 7, 649–673, 1954.

[166] G.I. Marchuk. *Some applications of splitting-up methods to the solution of mathematical physics problems* Aplikace Matematiky, 13, 103–132, 1968.

[167] G.I. Marchuk. *Splitting and alternating direction methods.* In Handbook of Numerical Analysis, P.G. Ciarlet, J.L. Lions (eds.), vol. 1. Elsevier Science Publishers, B. V.: North-Holland, 1990.

[168] R.I. McLachlan, G. Reinoult, and W. Quispel. *Splitting methods.* Acta Numerica, 11, 341–434, 2002.

[169] U. Miekkala and O. Nevanlinna. *Convergence of dynamic iteration methods for initial value problems.* SIAM Journal on Scientific and Statistical Computing, 8(4), 459–482, 1987.

[170] N. Morosoff. *Plasma Deposition, Treatment and Etching of Polymers.* R. d'Agostino ed., Acad. Press, First Edition, 1990.

[171] F. Neri. *Lie algebras and canonical integration.* University of Arizone, Department of Physics, Technical Report, 25 pages, 1987.

[172] N. Neuss. *A new sparse matrix storage methods for adaptive solving of large systems of reaction-diffusion-transport equations.* In Keil et. al., editor, Scientific computing in chemical Engineering II, Springer-Verlag Berlin-Heidelberg-New York, 175–182, 1999.

[173] O. Nevanlinna. *Remarks on Picard-Lindelöf Iteration, Part I.* BIT, 29, 328–346, 1989.

[174] O. Nevanlinna. *Remarks on Picard-Lindelöf Iteration, Part II.* BIT, 29, 535–562, 1989.

[175] O. Nevanlinna. *Linear Acceleration of Picard-Lindelöf.* Numerische Mathematik, 57, 147–156, 1990.

[176] X.B.. Nie, S.Y. Chen, W.N.E., and M.O. Robbins. *A continuum and molecular dynamics hybrid method for micro- and nano-fluid flow.* J. Fluid Mech., 500, 55–64, 2004.

[177] Web Site: *OFELI: http://ofeli.sourceforge.net/.*

[178] M. Ohlberg. *A Posteriori Error Estimates for Vertex Centered finite Volume Apprximations of Convection-Diffusion-Reaction Equations.* Preprints 12/2000, Mathematische Fakultät, Freiburg, May 2000.

[179] M. Ohring. *Materials Science of Thin Films.* Academic Press, San Diego, New York, Boston, London, second edition, 2002.

[180] A. Pazy. *Semigroups of Linear Operators and Applications to Partial Differential Equations.* Applied Mathematical Sciences, no. 44, Springer, Berlin, 1983.

[181] J. Prüss. *Maximal regularity for evolution equations in Lp-spaces.* Conf. Sem. Mat. Univ. Bari, 285, 1-39, 2003.

[182] A. Quarteroni and A. Valli. *Numerical Approximation of Partial Differential Equations* Springer Series in Computational Mathematics, Springer-Verlag Berlin-Heidelberg-New York, 1997.

[183] A. Quarteroni and A. Valli. *Domain Decomposition Methods for Partial Differential Equations* Series: Numerical Mathematics and Scientific Computation, Clarendon Press, Oxford, 1999.

[184] H. Rouch, M. Pons, A. Benezech, J.N. Barbier, C. Bernard, and R. Madar. *Modelling of CVD reactors: Thermochemical and mass transport approachesfor $Si_{1-x}Ge_x$ deposition.* Journal de Physique IV, 3, 17–23, 1993.

[185] H. Rouch. MOCVD Research Reactor Simulation. Proceedings of the COMSOL Users Conference 2006 Paris, Paris, France, 2006.

[186] M. Rumpf and A. Wierse. *GRAPE, Eine interaktive Umgebung für Visualisierung und Numerik.* Informatik, Forschung und Entwicklung, 1990.

[187] Y. Saad. *Analysis of some Krylov subspace approximation to the matrix exponential operator.* SIAM J. Numer. Anal., 29(1), 209–228, 1992.

[188] Y. Saad. *Iterative Methods for Sparse Linear Systems.* SIAM publications, 2nd edition, Philadelphia, 2003.

[189] H. Schmidt, P. Buchner, A. Datz, K. Dennerlein, S. Lang, and M. Waidhas. *Low-cost air-cooled PEFC stacks.* Journal of Power Sources, 105, 243–249, 2002.

[190] T.K. Senega and R.P. Brinkmann. *A multi-component transport model for non-equilibrium low-temperature low-pressure plasmas.* J. Phys. D: Appl.Phys., 39, 1606–1618, 2006.

[191] Q. Sheng. *Solving linear partial differential equations by exponential splitting.* IMA Journal of Numer. Analysis, 9, 199–212, 1989.

[192] B. Sportisse. *An analysis of operator splitting techniques in the stiff case.* Journal of Computational Physics, 161, 140–168, 2000.

[193] R. Steijl and G.N. Barakos. *Coupled Navier-Stokes - Molecular dynamics simulations using a multi-physics flow simulation framework.* International Journal for Numerical Methods in Fluids, John Wiley, 62(10), 1081–1106, 2009.

[194] G. Strang. *On the construction and comparision of difference schemes.* SIAM J. Numer. Anal., 5, 506–517, 1968.

[195] M. Suzuki. *General theory of fractal path-integrals with applications to many-body theories and statistical physics.* J. Math. Phys., 32(2), 400–407, 1991.

[196] A.-K. Tornberg and B. Engquist. *Numerical approximations of singular source terms in differential equations.* J. Comput. Phys., 200, 462–488, 2003.

[197] L. Verlet. *Computer experiments of classical fluids. I. Thermodynamical properties of Lennard-Jones molecules.* Physical Review, 159, 98-103, 1967.

[198] J.G. Verwer and B. Sportisse. *A Note on Operator Splitting in a Stiff Linear Case.* MAS-R9830, ISSN 1386-3703, 1998.

[199] S. Vandewalle. *Parallel Multigrid Waveform Relaxation for Parabolic Problems.* Teubner Skripten zur Numerik, B.G. Teubner Stuttgart, 1993.

[200] J. Waldén. *On the approximation of singular source terms in differential equations.* Numer. Meth. Part. D E, 15, 503–520, 1999.

[201] K. Yoshida. *Functional Analysis.* Classics in Mathematics, Springer-Verlag, Berlin-Heidelberg-New York, 1980.

[202] H. Yoshida. *Construction of higher order symplectic integrators.* Physics Letters A, 150(5–7), 262-268, 1990.

[203] Y. Zeng, Ch. Tian, and J. Liu. *Convection-diffusion derived gradient films on porous substrates and their microstructural characteristics.* Journal of Materials Science, 42(7), 2387–2392, 2007.

[204] Z. Zlatev. *Computer Treatment of Large Air Pollution Models.* Kluwer Academic Publishers, 1995.

Index

Printed and bound by CPI Group (UK) Ltd, Croydon, CR0 4YY

25/10/2024

01779207-0001